HALOCARBONS:
Environmental Effects of Chlorofluoromethane Release

Committee on Impacts of Stratospheric Change
Assembly of Mathematical and Physical Sciences
National Research Council

National Academy of Sciences
Washington, D.C.   1976

NOTICE: The project that is the subject of this report was approved by the Governing Board of the National Research Council, whose members are drawn from the Councils of the National Academy of Sciences, the National Academy of Engineering, and the Institute of Medicine. The members of the Committee responsible for the report were chosen for their special competences and with regard for appropriate balance.

This report has been reviewed by a group other than the authors according to procedures approved by a Report Review Committee consisting of members of the National Academy of Sciences, the National Academy of Engineering, and the Institute of Medicine.

Library of Congress Catalog Card Number 76-46750
International Standard Book Number 0-309-02529-X

Photocomposed on a UNIX time sharing system at
Bell Laboratories, Murray Hill, New Jersey 07974.

Available from:
Printing and Publishing Office, National Academy of Sciences
2101 Constitution Avenue, N.W., Washington, D.C. 20418

September 10, 1976

The Honorable H. Guyford Stever
Director
Office of Science and Technology Policy
Executive Office of the President
Washington, D.C. 20500

Dear Dr. Stever:

I am pleased to present the report "Halocarbons: Environmental Effects of Chlorofluoromethane Release." This report was prepared by the Committee on Impacts of Stratospheric Change, a committee of the Assembly of Mathematical and Physical Sciences of the National Research Council. The committee and study projects were chaired by Professor John W. Tukey of Princeton University and Bell Laboratories; the report is based in considerable measure on the findings presented to the committee in a report, also attached, of the Panel on Atmospheric Chemistry, chaired by Professor H. S. Gutowsky of the University of Illinois. The study was sponsored jointly by the National Science Foundation, National Oceanic and Atmospheric Administration, National Aeronautics and Space Administration and Environmental Protection Agency.

The report sets out what is known and, as best it can be judged, with what amount of uncertainty it is known. The report concludes that the selective regulation of chlorofluoromethane uses and releases is almost certain to be necessary at some time and to some degree. Neither the needed timing nor the needed degree can reasonably be specified today because of remaining uncertainties. However, measurement programs now under way promise to reduce these uncertainties quite considerably in the near future. The prospect for narrowing uncertainty and the finding that the rate of ozone reduction is relatively small engender in the committee the conclusion that a one- or two-year delay in actual implementation of a ban or regulation would not be unreasonable. Meanwhile, so that the government may be positioned to function in these regards, it seems highly desirable that the appropriate statutory basis for regulation be enacted at this time.

The uncertainties yet remaining suggest that a reexaminaton of these circumstances take place at periodic intervals. The National Research Council would be pleased to continue their cooperation in following these developments.

Finally, may I utilize this opportunity publicly to express to Dr. Tukey, Dr. Gutowsky and their colleagues our deep appreciation for their tireless and devoted efforts.

Sincerely yours,

PHILIP HANDLER
President
National Academy of Sciences

cc: Acting Director,
National Science Foundation
Administrator,
National Aeronautics and Space
Administration
Administrator,
National Oceanic and Atmospheric
Administration
Administrator,
Environmental Protection Agency
Chairman,
Council on Environmental Quality

Committee on Impacts of Stratospheric Change

Preface

In recent years it has become increasingly clear that human activities can affect the earth's ozone shield and that changes in the efficacy of this shield can have significant effects upon life on earth.

Awareness of the possible extent of such problems came first from concerns about the potential environmental impact of supersonic transports. These concerns led in 1972, in response to a request from the Department of Transportation (DOT), to the establishment of the Climatic Impact Committee—the precursor of the present Committee on Impacts of Stratospheric Change (CISC)—as a multidisciplinary committee within the National Research Council. In April 1975, the Committee completed its assessment of the climatic and biological impact of a projected fleet of high-altitude aircraft. The Committee's findings were released in the form of a report, Environmental Impact of Stratospheric Flight: Biological and Climatic Effects of Aircraft Emissions in the Stratosphere (National Academy of Sciences, Washington, D.C., 1975).

The Climatic Impact Committee had at its disposal the extensive results of the major Climatic Impact Assessment Program conducted by the DOT between 1971 and 1974 (in which some 350 scientists participated).

When attention was focused on the potential threat to the ozone posed by the use and release into the atmosphere of chlorofluoromethanes, a panel of the CISC was established in April 1975 to study the question of whether the chlorofluoromethanes would destroy ozone, and if so, what the magnitude of the effect would be and what uncertainties were associated with such a prediction. The panel was also asked to identify critical research needs to reduce the uncertainties.

The CISC meanwhile considered the question of biological and climatic effects of ozone reduction and the appropriate policy consequences of both our present knowledge and the knowledge we are likely to have in the future. In this effort the Committee was fortunate that it could draw, in part, upon the very recent work described in the stratospheric flight study as well as the wealth of information contained in the DOT's Climatic Impact Assessment Program Monographs.

The Committee hopes that the emphasis, in the present report, on the appropriateness, when and if our knowledge of the impact of chlorofluoromethane releases calls for control actions, of measures that selectively limit chlorofluoromethane uses, rather than ban the use of chlorofluoromethanes altogether, will provide a useful example of a regulatory attitude that, we are sure, will be frequently needed in the decades ahead.

The CISC is deeply grateful to the members of the Panel on Atmospheric Chemistry, whose exhaustive study of the potential impacts of chlorofluoromethanes on stratospheric ozone constitutes one crucial foundation of the present report.

The Committee appreciates the support provided by the National Science Foundation, the National Aeronautics and Space Administration, the Environmental Protection Agency, the National Oceanic and Atmospheric Administration, and the Federal Aviation Administration. We particularly appreciate the contributions and information provided by a wide range of people from universities, government, and industry to this effort. The problem of assessing a potential threat to environment and then evaluating the methods to deal with it will become increasingly important in the future. We hope that future groups will enjoy the cooperation and assistance that we have had in this effort.

John W. Tukey, *Chairman*
Committee on Impacts of Stratospheric Change

Contents

ix

Executive Summary

This summary is divided into Findings, Conclusions, Recommendations, the Problem of Regulation, and Background. A fuller understanding of the Findings and Recommendations will usually require a reading of the Background. (References to chapters and appendixes related to individual findings are also included; these can further supplement the Background.)

Two chlorofluoromethanes, F-11 and F-12, have been produced in increasing amounts and used, worldwide, for a diversity of purposes, including spray-can propellants, working fluids for air conditioners and refrigerators, and agents for producing certain solid plastic foams. In this report, CFMs will refer specifically to these two compounds (and will not include other chlorofluoromethanes).

In June 1974, the Federal Task Force on Inadvertent Modification of the Stratosphere (IMOS, 1975) reported

The task force has concluded that fluorocarbon releases to the environment are a legitimate cause for concern. Moreover, unless new scientific evidence is found to remove the cause for concern, it would seem necessary to restrict uses of fluorocarbons-11 and -12 to replacement of fluids in existing refrigeration and air-conditioning equipment and to closed recyled systems or other uses not involving release to the atmosphere.

The National Academy of Sciences is currently conducting an in-depth scientific study of man-made impacts on the stratosphere and will report in less than a year. If the National Academy of Sciences confirms the current task force assessment, it is recommended that the Federal regulatory agencies initiate rulemaking procedures for implementing regulations to restrict fluorocarbon uses. Such restrictions could reasonably be effective by January 1978—a date that, given the concern expressed now, should allow time for consideration of further research results and for the affected industries and consumers to initiate adjustments.

1

In preparing this report, the Committee on Impacts of Stratospheric Change has felt that its responsibilities, as a body of scientists charged with preparing an honest assessment of the evidence, included doing the best it could to distinguish among three possibilities:

1. The scientific evidence for the proposed consequences of CFM releases is demonstrably sound and gives reasonably certain answers;
2. The scientific evidence for the proposed consequences of CFM releases when carefully examined is coherent but leaves us with a substantial range of uncertainty;
3. The scientific evidence for the proposed consequences of CFM releases contained fundamental flaws that are not immediately repairable.

To try to distinguish only between the first and third of these would be to neglect a real, and possibly important, contingency. As discussed briefly just after our recommendations, the scientific picture has changed considerably during the 14 months since the issuance of the IMOS report.

In thinking about CFMs, it is important to recognize that they are a world-wide problem. Whatever is released is, before long, mixed throughout the atmosphere. Rather less than half of the 1975 uses and releases came from the United States, rather more than half from other countries. Total cessation of uses in the United States would not quite halve uses and releases worldwide. Halving worldwide releases could more reasonably happen by halving uses and releases in all heavily using countries.

Just as the United States took the lead in pioneering the varied uses of CFMs, so too if there is to be worldwide control of CFM uses and releases, the United States will have to take the lead here also, when and if this is appropriate. The Congress and the Executive Branch will thus need to consider the effect of their decisions in encouraging similar actions by other countries, whose own decision will affect more than half of present releases and thus more than half of any change in impact of CFMs on the United States.

FINDINGS

EFFECTS OF CFM RELEASE

We find that

(A) The accumulation of CFMs in the atmosphere, at all levels, increases the absorption and emission of infrared radiation. This retards heat losses from the earth and thus affects the earth's temperature and climate. The amount of change in infrared absorption and emission is well known, but both the amount and details of the further effects on the earth's climate are uncertain. This CFM effect is inevitably combined with the effect due to increased CO_2 and acts in the same direction. (See also Chapter 6.)

(B) CFMs, after release at the surface of the earth, mix with the atmosphere and rise slowly into the stratosphere, where they are decomposed by the sun's ultraviolet radiation. Chlorine atoms (Cl) and chlorine oxide (ClO), produced directly or indirectly by this decomposition, then react to remove ozone (catalytically), reducing the total amount of ozone and somewhat shifting the distribution of ozone toward lower altitudes. As a consequence (see also Chapter 5):

—More biologically active ultraviolet (DUV) reaches the earth's surface.
—The temperature distribution in the stratosphere is somewhat altered.

The reductions in ozone take place over a long time, individual release of CFMs having effects spread over decades.

(C) The extent of ozone reduction attributable to CFMs has not been measured. Because of the natural variations in the amount of ozone above us, much larger than any ozone reduction so far caused by CFMs, direct verification of CFM effects will not be feasible for at least several years. (See also Chapter 5.)

(D) At the moment, the ozone reduction and consequent DUV increase corresponding to a given CFM release is uncertain by a large factor. Continued release at the 1973 level, the usual example, is calculated to give an ultimate reduction in ozone of about 7 percent, where "about 7 percent" is relatively certain to be between 2 percent and 20 percent. This range does not allow for possible inadequacies of the bases of the calculation. Three of the possible kinds of inadequacies may be cited as examples: (1) essential chemical reactions not so far recognized as such, (2) the possibility of unexpected effects of tropospheric sinks (many possible sinks have been studied carefully), (3) possible important inadequacies in the one-dimensional transport models. (See also Chapter 5.)

(E) Continued CFM release at 1973 levels could by the year 2000 produce about half of the direct climatic effect caused by CO_2 increase over the same period, although the magnitude of both effects on climate is less certain. Thus, the CFM effect may well deserve serious concern. (See also Chapter 6, especially Table 5.)

(F) In our present state of knowledge, it would be imprudent to accept increasing CFM use and release, either in the United States or worldwide. (Recent reductions in CFM releases are ascribed by some to economic conditions and by others to consumer pressure, real or anticipated.) (See also Chapters 5, 6, and 7.)

However, we also find that

(G) Advances in our knowledge of climate mechanisms over the next two years will improve our assessment of climatic effects due to CFMs (both through ozone reduction and displacement and through infrared absorption), but these advances cannot be expected to make our assessment of the climatic effects as precise as our assessments of ozone reduction and DUV increase. (See also Chapter 6.)

(H) The range of uncertainty about the amounts of ozone reduction and DUV increase consequent on a given CFM release pattern can be considerably reduced during the next two years; new stratospheric measurements (particularly those from the substantial program supported by the National Aeronautics and Space Administration), measurements of atmospheric CFMs, and improved laboratory measurements will contribute to this. More importantly, the possibility of unexpected inadequacies in the basis of our calculations will be greatly reduced by more extensive and better measurements. (See also Chapter 5.)

(I) Many other improvements in our knowledge can be attained over the next five to ten years, if we push hard to do this (see pages 1–17), but others will take still longer to attain. (See, for example, Chapter 7.)

HOW SLOWLY DO THINGS HAPPEN?

We find that

(J) If CFM uses and releases were to continue at a constant rate, the ozone reduction and consequent DUV increase would gradually flatten out, approaching a steady state. To reach half of this value would take roughly 50 years. In particular, if constant CFM releases at the 1973 rate are to give 7 percent ultimate reduction of ozone, this reduction will initially increase at about 0.1 percent a year, reaching 3.5 percent after roughly 50 years. (See also Chapter 5.)

(K) If the rate of CFM release, after continuing at a constant rate, were drastically reduced at any time in the next decade, say halved or eliminated, and then continued at the drastically reduced rate, ozone reduction and consequent DUV increase would continue to increase for at least a decade after the drastic reduction. It would then decrease, if releases had been nearly eliminated, by roughly 1/70 of its current value each year, taking roughly 50 years to fall back to half its peak value. (See also Chapter 5.)

(L) If CFM use and release were to continue at a constant rate, the amount of direct climatic effect would also flatten out, approaching a steady state, again reaching half of this value in about 50 years. The increase of infrared absorption and emission would similarly reach half of its ultimate value in about 50 years. Resulting climatic effects might be further delayed because of slowness in response in the climatic mechanism. (See also Chapter 6.)

(M) If the rate of CFM use and release were nearly eliminated at some date, the increase in infrared absorption and emission would, by contrast, begin to decrease immediately, with any delays arising only from the climatic mechanism itself. It would then decrease by roughly 1/70 of itself each year taking roughly 50 years to reach half of the value in cutoff. (See also Chapter 6.)

WHAT ARE THE IMPACTS?

We find that

(N) The major effects of DUV increase due to ozone reduction could involve

—Increased incidence of malignant melanoma, a serious form of skin cancer frequently causing death, and thus an increase in mortality from this cause (see also Chapter 8 and Appendix A);
—Increased incidence of basal- and squamous-cell carcinomas, less serious but much more prevalent forms of skin cancer, rarely causing death but causing much expense and, occasionally, more or less serious disfigurement (see also Chapter 8);
—Effects on plants and animals of unknown magnitude (see also Chapter 7 and Appendix C).

Whether the first of these effects, melanoma increase, will occur is not firmly proven, but the evidence of its plausibility is now strong enough for it to be treated as a serious health hazard. The second effect, nonmelanoma increase, is relatively well established, and its amount reasonably assessable (see Climatic Impact Committee, 1975, pages 41–45). The third group of effects, action of DUV increases on plants and animals, is only beginning to be explored. For the present there is good reason for a strong concern to know more about this third group of effects, but, as yet, there is no clear indication of their seriousness.

(We are unlikely to make major strides in our knowledge of the connections, actual or potential, between DUV increase and any of these major effects during the next two years, although it is important to continue active work in each of these directions.)

(O) If the increased infrared absorption and emission due to the presence of CFMs in the atmosphere were to alter our climate by small amounts, the most important effects would be on agriculture, particularly through the boundaries of the regions in which particular crops can be grown effectively. (Other agriculatural effects are possible.) (See also Chapters 6 and 7.) [The influences of small climate changes on agricultural production are not easy to assess (see Climatic Impact Committee, 1975, pages 58–63), but the uncertainties here are less than those in the amount of climate change consequent on a given release of CFMs.]

WHAT ABOUT CFMs?

We find that

(P) Worldwide, CFM uses and releases grew by about 10 percent a year through 1974. They decreased by about 15 percent in 1975. Recently, about half of the uses and releases have come from the United States. Most, but not all, of the 1975 decrease came from decreased use in the United States. (See also Chapter 5.)

(Q) The various uses of CFMs are of very different magnitude and of very different importance to human life, including human health. Home refrigeration of food, at one extreme, is important to human health and accounts for less than 1 percent of all releases. CFM uses in aerosol sprays, at the other extreme, are mainly replaceable by other dispensing techniques or by other propellant substances, at some loss in convenience, efficiency, or safety, and amount to about three quarters of all releases. (Some aerosol uses, including some for medical purposes, deserve special consideration.) (See also Appendix D.)

WHAT ARE THE PENALTIES OF DELAY?

(R) When the time history of past releases is considered, and based upon an ultimate ozone reduction of 7 percent (central value of 2 percent to 20 percent range), whether a halving in CFM use and release were to take place in 1978 or in 1980 would alter the ozone reduction at any later date by no more than 1/6 percent (central value of a 1/18 percent to 1/2 percent range). The difference in ultimate ozone reduction, if uses and releases continued at the halved level in each case, would be less than 1/10 of a percent (central value of a 1/30 percent to 1/4 percent range) after at least half of the ultimate reduction is reached (after roughly 50 years). (See also Chapter 5.)

(S) Whether a halving of CFM use and release were to take place in 1978 or in 1980 would alter the total amount of CFMs in the atmosphere by no more than 10 percent of the amount now present—by no more than 10 percent of an amount whose climatic effects are probably undetectably small. (See also Chapter 6.)

CONCLUSIONS

Selective regulation of CFM uses and releases is almost certain to be necessary at some time and to some degree of completeness. Neither the needed timing nor the needed severity can be reasonably specified today. Costs of delay in decision are small, not more than a fraction of a percent change in ozone depletion for a couple of years' delay. Measurement programs now under way promise to reduce our uncertainties quite considerably in the near future.

RECOMMENDATIONS

Accordingly

1. As soon as the inadequacies in the bases of present calculations are significantly reduced, for which no more than 2 years need be allowed, and provided that ultimate ozone reductions of more than a few percent then remain a major possibility, we recommend undertaking selective regulation of the uses and releases of CFMs on the basis of ozone reduction.

2. We recommend that, as soon as appropriate legislative authority is in place, as well as every three to five years thereafter, our current knowledge of the importance and the certainty or uncertainty of the direct climate effect be reviewed, so that appropriate decisions can be taken about regulation of CFM uses and releases on the basis of this effect. In so doing, the effects of CFM increases should be considered in the light of the effects of CO_2 increases with which they are inevitably combined.

3. Whenever regulation is undertaken, we recommend that it should be selective, treating one use differently from another, both as to whether a particular use is to be excepted from regulation or not and as to the time allowed for compliance with regulation. (See also Chapter 4 and especially Appendix A.)

4. Legislative authority may not now be adequate among other things to (a) regulate the uses of CFMs selectively, (b) regulate the handling of CFMs (as in repairing auto air conditioners, for example), and (c) regulate CFMs on the basis of threats to plants and animals important to human life (either through DUV increase or climate changes) rather than on the basis of threats

to human health. We recommend that immediate steps be taken, first to determine what inadequacies in legislative authority exist and then to eliminate, through additional legislation, those that exist.

5. Since carefully informative labeling would allow consumers an opportunity to distinguish, for example, CFM-propelled aerosols from aerosols using other propellants, we recommend that legislation be enacted requiring labeling of all products containing the CFMs F-11 and F-12 and not intended to remain under seal during use. (Aerosol cans and refill containers for air conditioners and refrigerators would then require labels; automobiles and refrigerators themselves would not.) Labeling should in no sense be regarded as a substitute for regulation but rather as an aid to consumer self-restraint in the use of CFMs and to consumer preparation for possible later regulation.

6. In view of the present inadequacies in the bases of our calculations, in view of the reduction in these inadequacies promised by ongoing measurement programs, and in view of the small changes in ozone reductions following from a year or two delay, we wish to recommend against decision to regulate at this time. (See also The Problem of Regulation, below.)

7. When and if regulation is decided upon by the United States, similar action by other countries should be encouraged by whatever appropriate means are most likely to be effective.

We also make the following recommendations:

8. Since both further laboratory studies and, especially, well-enough planned measurements in the atmosphere, can do much over the next few years to improve the basis for well-chosen regulation, we recommend that laboratory studies and atmospheric measurements should be given an appropriately high priority. (See also Chapter 2.)

9. Since there are at least two important areas,
   (a) The mechanisms of climate determination and climatic change,
   (b) The effects of increased (or decreased) biologically active ultraviolet radiation on plants and animals,
where we still lack an adequate scientific foundation, and since adequate foundations cannot be constructed by short-term "crash" programs, longer-term research programs, extending over several years and guided by the consensus of the best scientific minds available, should be established and adequately funded in each such area as a matter of urgency. (See also Chapter 2.)

10. Since learning to identify population groups with drastically higher susceptibility to melanoma (and to other skin cancers) will greatly increase

the efficiency and effectiveness with which individuals can be taught to protect themselves, it is urgent to undertake a program of learning better to identify such susceptible groups. (See also Chapter 2 and Appendix E.)

11. Information about the relative releases of CFMs from different uses would be so essential, if and when control of CFM release becomes appropriate, that vigorous efforts should be made to provide such information on a continuing basis.

12. Since ultraviolet-induced skin cancer will continue to present a serious health hazard, we need to study possible preventive medicine actions carefully, without regard to the effectiveness of CFMs in reducing ozone or decisions about their regulation. (See also Chapter 2 and Appendix E.)

THE PROBLEM OF REGULATION

This report makes two things clear. The impact on the world of waiting a couple of years before deciding whether or not to regulate the uses and releases of F-11 and F-12 is small although we are uncertain just how small. The impact on industry of a ban on uses of F-11 or F-12 in most types of spray cans would be appreciable. Against a background of a possible, although very small, change in world climate, however, the industrial impact does not loom large.

This Committee in meeting its responsibilities to assess what is scientifically known has focused on uncertainty about the adequacy of the bases of our calculation and recommended a brief delay before decision. Some scientists, emphasizing the possible critical importance of even small effects on climate and the relative unimportance of many spray-can uses, might well urge immediate decision to regulate, although authority to regulate on the basis of climate effects seems still to be lacking.

The report sets out what is known and, as best as this can be judged, with what amount of uncertainty it is known. The choice of when to make decisions about regulation is a political one in the highest sense of that word. The ultimate balance—between increased impact on industry and on spray-can uses, on the one hand, and possibly climatic impacts and more certain skin cancer increases, both very small for a short delay, on the other—has inevitably to be made by those who decide for the whole of each country concerned, in the United States by its Congress and President.

The report stresses the fundamental importance, clearly illustrated in this case but much more widely applicable, of conducting such regulation use by use. It is not sufficient to label a substance "good" or "bad." We often need, as we do here, to look use by use to see how important each is to human living—often, as here, to human health—and compare this with the size of the unfavorable impacts from that use. To begin to do this is not easy, but our world is complex enough to force us to face such difficulties more and more frequently.

Having laid open the facts as best it can and stressed the fundamental importance of regulation use by use, the scientific community as represented through the National Academies of Sciences and Engineering and the National Research Council, can, we believe, properly leave decisions about timing, in this country, to the Congress of the United States. (Individual scientists and engineers will no doubt wish to participate in the debate from a variety of points of view.)

BACKGROUND

INTRODUCTION

By early 1975, scientific knowledge and insight about the consequences of CFM release had developed to the point where the IMOS (1975) report could focus concern on their possible or probable effects on the amount of ozone spread through the upper parts of our atmosphere. Significant reductions in ozone would produce significant increases in the amounts of relatively short-wavelength ultraviolet light (uv-B) reaching the earth's surface. This uv-B is biologically highly active, producing in particular, both immediate effects, such as sunburn, and long-term effects, such as skin cancer (and effects of so far almost unknown magnitude on plants and animals).

Two relevant scientific discoveries were made during the past year:

—Ramanathan's (1975) observation that F-11 and F-12 are sufficiently effective absorbers of infrared light (near 10 mm) to have a potentially significant effect on the earth's heat balance.
—The unexpected relative insensitivity of the compound chlorine nitrate ($ClONO_2$) to decomposition by ultraviolet light determined in measurements by Rowland *et al.* (1976).

The impact of these two discoveries, integrated into the total picture as well as has been possible with the limited time available, leaves us with a need to assess the consequences of CFM releases in at least three directions:

1. The change in the amount of DUV reaching the earth, brought about by changes in the total amount of ozone overhead.
2. The decrease of the transparency of the atmosphere in the infrared near 10 mm (and thence, to an as yet inadequately understood degree, to changes in the earth's climate).

3. The change in the patterns of heating in the stratosphere caused by changes in the distribution of ozone with height.

The first effect, increased DUV reaching the earth, produces increases in human skin cancer and may have other biological effects. The second and third effects contribute to changes in climate.

Only the second effect comes directly from the CFMs themselves and is produced in all parts of the atmosphere. Of the uncertainties in assessing the impacts of CFMs, only the existence of so-far undemonstrated major mechanisms for the destruction or trapping of CFMs would influence the extent of this effect.

The first and third effects come from the decomposition of CFMs by ultraviolet sunlight in the upper parts of the stratosphere and the consequent reductions in ozone by chemical chain reactions involving these decomposition products. The extent of these reductions, for a given amount of CFMs decomposed, is sensitive to a variety of incompletely known constants, reactions, and phenomena. The range of uncertainty from these, when carefully assessed, is relatively wide—the smallest likely effects being about one tenth of the largest—and there are possibilities for future surprises. (A recent surprise involved laboratory assessments of the rates of formation and decomposition of chlorine nitrate, which were actively studied in the second quarter of 1976. Different results for these rates would have led to a distinctly different assessment of the importance of concerns 1 and 3.)

CONCLUSIONS OF THE PANEL ON ATMOSPHERIC CHEMISTRY

All the evidence examined by the Panel on Atmospheric Chemistry indicates that if the releases of F-11 and F-12 continue at present rates for a long time, they will cause a reduction in the amount of stratospheric ozone by something like 7 percent. Using the presently best founded bases for calculation, there is at least a tenfold range of uncertainty about this central value, i.e., the reduction may be as little as 2 percent or as much as 20 percent, using what are believed to be roughly 95 percent confidence limits.* Further allowance for uncertainty would be required if some so far unallowed for contingency were to materialize, thus changing the bases for calculations. Those Panel members who are members of this Committee, as a result of the intensive study summarized in the Panel report, are seriously concerned about the possibility of such contingencies.)

---

*See comment at the close of Chapter 5 for a discussion of how these limits arise and their appropriateness.

At constant CFM releases, 40 to 50 years will be required to attain half of whatever steady-state ozone reduction will occur. Even if CFM releases were terminated within the next several years, the ozone reduction would continue to rise for about a decade, reaching a peak at least one and one-half times as large as the ozone reduction at the time of termination. Furthermore, 50 to 75 years would be needed for the ozone reduction to return to a value one half of the peak reduction.

In the Panel's consideration of the major sources of uncertainty in its predictions, numerical estimates were made of the contributions from several sources: uncertainties in the measurements of seven of the reaction rate constants (a fivefold range); approximations of the one-dimensional calculations used in the predictions (a threefold range); treatment of photochemical processes and measurements of concentrations of natural stratospheric species (a twofold range for each). In addition, it should be noted that the predictions are based on a reaction scheme that includes the formation of chlorine nitrate ($ClONO_2$), the presence of which has not yet been detected in the stratosphere. If $ClONO_2$ proves to be less important than indicated by the present data, the ozone reduction could be larger than the stated values by a factor of up to 1.85 (giving a value of about 13 percent, uncertain by an overall factor of about 9).

Other less tangible aspects that affect the accuracy of the predictions are the completeness of the reaction scheme employed for the stratospheric chemistry and the possible occurrence of a variety of feedback and inactive removal processes. A considerable number of such possibilities have been analyzed in both areas. There is limited evidence supporting oceanic removal of F-11 at a rate that might decrease the predictions of ozone reduction by a factor of one fifth (20 percent) from the values they would otherwise be. (The 7 percent quoted above is a compromise between 7.5 percent with no such sink and 6 percent with one.) An effort was made to place a meaningful upper limit on the amount of CFM that might be removed by all such inactive removal processes by comparing the amount of CFM released with the amount observed to be in the atmosphere (by making a materials balance). However, the atmospheric CFM measurements are still too sparse and inaccurate for such results to be meaningful.

It seems likely that the uncertainties in the bases of calculation can be reduced materially during the next year or two. Infrared measurements now under way should establish the role of $ClONO_2$ in the stratosphere. Measurements of ClO and Cl concentrations as well as of those other species central to the stratospheric chemistry of the CFMs should help to determine the completeness of the reaction scheme employed in the calculations. An accurate materials balance, if feasible, would reduce the likelihood of unidentified sinks of major consequence. Also within the next couple of years, experimental errors in rate constants are likely to be reduced materially, as well as uncertainties in photochemical processes and in the concentrations of natural species.

Continuation of CFM releases at their present (static) levels, while we wait for improvements in our ability to determine the ozone reduction caused by the CFMs, will increase both the eventual peak ozone reduction and the aggregate amount of ozone reductions (accumulated over all relevant time) above the values that would have had to be faced if release were curtailed at once. Each year of release, compared with complete cessation, will increase both the peak reduction by about 0.07 percent (central value of a 0.02 to 0.2 percent range) and the total amount of reduction (accumulated over time) by one tenth. (The peak effect on infrared absorption would also be increased by one tenth.) If we compare release at 1973 rates with release at half of that rate, each year of delay will have about half as much effect on ozone reduction as just stated for comparison with complete cessation. A resumption of exponential growth would, of course, give annual increments of increasing size.

THE MELANOMA PROBLEM

The occurrence of the most serious form of skin cancer, malignant melanoma, may well be related to exposures to solar ultraviolet radiation. The relationship, which this Committee believes to be likely but yet not completely proven, is not a simple one.

The less serious (and much more prevalent) forms of skin cancer, basal- and squamous-cell carcinomas, are quite clearly associated with uv exposure. (See Climatic Impact Committee, 1975, pages 36–41, for discussion and background.)

The evidence concerning melanoma skin cancer is discussed in more detail in Chapter 8, and briefly summarized, in comparison with that for the nonmelanoma skin cancer in Table 1.

CLIMATE EFFECTS

We attach serious concern to the possible effects on climate. For the present, however, it is not feasible to attach numbers to the effects to be expected—even numbers as ill determined as those we have assigned to the ozone reduction. The direct climate effect, which seems most likely to be serious, is immediate (is not delayed for the CFMs to reach the stratosphere) and shows no overshoot (the effects of curbed release show more rapidly). It is probably not a concern at present levels of CFM accumulation. Thus, delay in decision is even more reasonable, as far as direct climate effects go, than for the effects of ozone reduction.

TABLE 1   Summary of Kinds (and Strength) of Evidence Relating Skin Cancer (Nonmelanoma and Melanoma) to Ultraviolet Exposure

	Nonmelanoma	Melanoma
Evidence from Incidence or Mortality in Groups		
Latitude dependence	Strong[1]	Strong[1,2]
Low incidence in more pigmented peoples[a]	Strong[3,4]	Strong[4]
Occupational differences	Strong	[b]
High incidence among persons more susceptible to sunburn	Weak[5]	Moderate[6]
Distribution over Body		
Habitually exposed areas (hands, face, etc.)	Very frequent[5]	Less frequent[7]
Relatively exposed areas[c] (trunk, legs of females)	Infrequent[5]	Frequent[7]
Rarely exposed areas (e.g., those covered by conventional bathing suits)	Infrequent[5]	Infrequent[7]
Behavior in Animals		
Uv action spectrum	uv-B[8]	Not known

[a]Blacks, Asiatics, American Indians, East Indians, etc.

[b]Substantial differences not consistent with total exposure as the controlling effect but (see Chapter 8) consistent with more complex possible relations.

[c]Clinical experience with cases of sunburn indicates that considerable uv-B reaches such areas.

[1]CIC report, Figure C.2, p. 184.

[2]J. Scotto, *J. Nat. Cancer Inst.* 56(3):489-491 (1976).

[3]H. F. Dorn, *Public Health Rep.* 59:33, 65, 97 (1944).

[4]J. D. Fleming et al., *Cancer* 35(3):600-605 (1975).

[5]F.F. Urbach, *Advances in Biology of Skin: Carcinogenesis, Vol. VII.* N. Montagna and R. L. Dobson, eds. Pergamon Press, New York, 1966, p. 195 (no controlled data).

[6]Melanoma Clinical Cooperative Group (U.S.A.). 343 patients, 150 matched controls (controlled data).

[7]Melanoma Clinical Cooperative Group (U.S.A.). 534 patients.

[8]R. G. Freeman et al., *Int. J. Dermatol.* 9:232-235 (1970).

It remains highly important for us to understand what climate modifications will follow changes in infrared absorption and emission, not only because of CFM accumulation but also because of the accumulation of carbon dioxide (CO_2), the main consequence of gaining energy from the burning of oil and coal.

INCREASE OF KNOWLEDGE

Two years hence (1978), we could, if we pressed forward vigorously:

—Significantly reduce the possibility of subsequently finding an unidentified factor that has a major effect on predictions of ozone reduction by the CFMs. Also, we should be able to reduce the overall (identified) uncertainty of the predictions from a tenfold to a fourfold or a fivefold range.
—Have begun to use more sophisticated models of the stratospheric transport and chemistry to estimate total ozone and the distribution of ozone with height.
—Have made useful inquiries into the climatic effects (and their consequences) to be expected from the absorption of infrared light by the CFMs.
—Have begun to understand the likely consequences to climate of redistributions of ozone in height.

We might hope, in 5 to 10 years, again if we pressed forward vigorously, to clarify, at least in significant part:

—The detailed climatic consequences of the absorption of infrared radiation by the CFMs.
—The consequences of possible climatic changes for agricultural productivity, for energy and water resource planning, and for accompanying changes in polar ice sheets and sea level.
—Consequences of changes in the altitude distribution of ozone, including the consequences to the temperature structure of the stratosphere and the consequences to climate.
—The probable effects of ozone decrease, through ultraviolet increase, on the major crops and forests of the world. (It will be necessary first to untangle the effects on the major processes of plant growth and decay. This will take more than a few years.)
—The quantitative aspects of the apparent relation of human exposure to the ultraviolet components of sunlight to skin cancer.

DISTRIBUTION OF USES

The general distribution of releases by general categories of use of CFMs in 1975 is shown in Table 2. The uncertainties in the figures given vary somewhat, but the orders of magnitude are believed correct. A somewhat more detailed breakdown and a description of sources and assumptions will be found in Appendix D.

TABLE 2 Estimated 1975 Worldwide Releases of F-11 and F-12[a]

	Million of pounds	Amount (%)
Aerosols		
Antiperspirant/deodorants	458	31
Hair care	402	27
Other personal care	75	
Household	69	
Insecticides	33	
Miscellaneous	78	
(SUBTOTAL)	(1115)	(74)
Cooling		
Vehicle air conditioners	90	6
Building cooling	43	
Food and beverages	39	3
Home refrigerators and freezers	6	0.4
Miscellaneous air conditioners/ refrigerators	27	
(SUBTOTAL)	(204)	(14)
Foams		
Open-cell foams	100	
Closed-cell foams	76	
(SUBTOTAL)	(176)	(12)
TOTAL	1495	100

[a]Total releases; breakdown between F-11 and F-12 unavailable. See Appendix D for further details. Note that F-11 (CFCl$_3$) is 77 percent chlorine by weight, while F-12 (CF$_2$Cl$_2$) is 59 percent, so that if F-11 could be replaced by an equal weight of F-12, unfortunately not often feasible, the amount of ozone reduction would be decreased.

SCENARIOS

Worldwide, the production (and subsequent release) of the CFMs (F-11 and F-12) grew at about 10 percent a year through 1974. Over the last 2 years, there has been a decline in worldwide production amounting to about 15 percent, more than 20 percent in the United States and 4 or 5 percent in the rest of the world. (Current production is about half in the United States and half outside the United States.) The consequences of some alternative scenarios (possible patterns of future releases) are as shown in Table 3, which assumes no tropospheric sink.

If CFM use, both in the United States and elsewhere, returns to the pattern of growth established during the 1960's and early 1970's, it is clear that the amount of CFMs in the atmosphere will increase severalfold by the year 2000 from the current value. (It would nearly triple if releases were to continue at the 1973 rate.)

TABLE 3 Relative Amounts of CFMs Remaining in the Atmosphere and Total Accumulated Releases According to a Few Simple Scenarios with Removal by Photolysis Only

Scenario	CFMs Remaining in Atmosphere at Year's End (Cumulative Total Released)		
	1980	1990	2000
1973 rate continuing $(1)^a$	$14^b(15)^c$	$21^b(25)^c$	$28^b(\text{et})^c$
1975 rate continuing (0.9)	13 (15)	20 (24)	26 (33)
1975 rate until 1979 thence 1/2 1975 rate^d (0.45)	13 (14)	15 (19)	18 (23)
1975 rate until 1979 cutoff (0)	12 (14)	11 (14)	9 (14)
1975 rate until 1978 thence 10 percent growth* *(1.1 in 1980, 2.8 in 1990, 7.3 in 2000)	13 (15)	30 (34)	74 (83)

^{a}Number in parentheses indicates release rate in units of the 1973 release rate.

^{b}Amount of CFMs remaining in the atmosphere as a multiple of the 1973 release rate.

^{c}Cumulative amount of CFMs released as a multiple of the 1973 release rate.

^{d}Halving would most likely involve action by several countries, including the United States.

RANGES OF UNCERTAINTY FOR PHYSICAL CONSEQUENCES

The part of the uncertainty about the impact of CFMs on total ozone that we can recognize and try to assign numbers to is large. The Panel on Atmospheric Chemistry's judgment of the consequences of continuing release at 1973 rates, for example, leads to an eventual ozone reduction that may be as small as 2 percent (which we think most informed persons would judge to be tolerable) or as large as 20 percent (which we think most informed persons would judge to be intolerable). This range has also to be broadened to allow for as yet undeterminable uncertainties.

Uncertainties about the impact of CFM accumulation on climate are more basic. Putting them into numbers here would probably be unprofitable. But if CFM releases continue to increase, it is only a matter of time before the consequences will be unacceptable. The uncertainty here is mainly about how soon this will happen—about what is the largest amount of CFMs that we can allow to accumulate in the atmosphere.

Uncertainties about the climatic consequences of changes in distribution of ozone with height are even less well defined.

TOLERABILITY

If the range of 2 percent to 20 percent ultimate reductions were firm, did not have to be further broadened to allow for still undetermined uncertainties, it would be quite possible—although not easy—to argue carefully and correctly that one extreme was tolerable, the other intolerable. In doing this, it would be necessary to consider such questions as (a) the size of natural variations in ozone and uv-B; (b) the number of deaths tolerated as consequences of other human activities; (c) the cost, to the nation or to the world, of reducing deaths by one means rather than another.

When we allow for uncertainties

—About possible contingencies affecting the bases of the calculations of ozone reduction,
—In the relation between increased DUV and increased melanoma deaths

our problem is much simpler. If the ultimate ozone reduction were to be appreciably less than 2 percent—which means appreciably less than 1 percent at any time in the next 50 years—and if the relation of DUV increase to melanoma deaths were appreciably weaker than we fear, the consequences of continuing CFM uses and releases at 1973 rates, although not desirable, would be tolerable. If on the other hand, the ultimate ozone reduction were even greater than 20 percent, the dangers to plants and animals as well as to humans, although still uncertain, would surely be intolerable.

The important point here is simple. What we know is sufficiently uncertain that continued releases of CFMs at 1973 rates:

—May be tolerable
—May be intolerable

If we did not have good reason to expect considerable improvements in our knowledge within a limited and foreseeable time, we would have no alternative but to try to assess the chance of intolerability and to make recommendations about the difficult decisions involved in dealing with an uncertain risk. Since brief delay should not have serious consequences, we have made recommendations 1, 2, and 6, calling for a strictly limited delay.

INTERNATIONAL ASPECTS

Sun-related skin cancers occur in white-skinned people. The population of Europe, excluding the Soviet Union, is rather more than twice that of the United States. However, this larger population is subject to less solar uv radiation (roughly one half) because it lies rather more than 10° further north. Thus, Europe's total susceptibility to CFM-related increases of skin cancer caused by solar uv radiation seems likely to be about the same as that of the United States. Adding Canada, the Soviet Union, and Australia increases the non-U.S. susceptibility somewhat above that for the United States. Since non-U.S. releases of CFMs are about equal to (in 1975 somewhat greater than) those from U.S. uses, the balance between releases and sensitivity is somewhere near equal. It is thus not seriously unfair to base discussion of melanoma tolerability on U.S. cases.

The situation with climate changes, which might bear on everyone in the world, perhaps more importantly toward the poles, is another matter. Any assessment of climatic impact must, following the lead of the report of the Climatic Impact Committee (1975), consider the effects all over the world.

ALTERNATIVE ROUTES TO HUMAN PROTECTION

Were it true, as seems likely, that only a small—and eventually identifiable—portion of our population is susceptible to melanoma, the ability to identify the susceptible persons would offer us a real opportunity to try to protect them by specific programs of education, including how to screen themselves for melanoma and what precautionary measures they should take. This might be a more effective way of dealing with melanoma than by avoiding ozone reduction, since complete avoidance of ozone reduction would leave us with, in the absence of public-health measures, more than 2000 deaths a year from melanoma, which it would be important to reduce. It is not yet a feasible way, however, since we cannot yet pick out susceptibles effectively enough.

The situation with other forms of skin cancer may be similar, although the group having to take protective measures may prove larger.

OVERALL SITUATION

The CFMs may well face us with a situation for which we will have to deal worldwide with control of a material whose use in small quantities is helpful to humanity, while its use in large quantities is unacceptable. It is perhaps confusing that our first major step into a new kind of regulation may turn out to be one for which the scientific parameters and quantitative assessments of consequences are still in considerable disarray. But it seems likely that this will occur frequently as more such problems arise in a world where interdependence will be increasingly forced by limitations of atmosphere, oceans, and natural resources—and not just by the interdependence of exchanges of goods and sources. Guidelines for the possible future control and regulation of CFM release are set down in Appendix A. They involve selective controls for each particular end use (and a discussion of the difficulties with a system of use-inhibiting taxation).

REFERENCES

Climatic Impact Committee. 1975. *Environmental Impact of Stratospheric Flight: Biological and Climatic Effects of Aircraft Emissions in the Stratosphere.* National Academy of Sciences, Washington, D.C.

IMOS. 1975. Report of the Federal Task Force on Inadvertant Modification of the Stratosphere. *Fluorocarbons and the Environment.* Council on Environmental Quality, Federal Council for Science and Technology.

Ramanathan, V. 1975. Greenhouse effect due to chlorofluorocarbons: climatic implications. *Science* 190:50–52.

Rowland, F. S., J. E. Spencer, and M. J. Molina. 1976. Stratospheric formation and photolysis of chlorine nitrates, $ClONO_2$. Preprint.

Recommendations on
Further Knowledge

ESTABLISHMENT OF A RESEARCH PROGRAM

Some important uncertainties about the likely consequences of continued CFM release on human affairs stem from fundamental gaps in our present knowledge. This chapter describes a number of points about which further understanding is required to reduce these uncertainties. Taken together, they establish the need for a systematic and sustained research program under federal support.

It is easy to recommend expenditures of federal funds. It is somewhat harder to assure that they will contribute effectively to the intended goals. Where the goals are an acquisition of knowledge—always a process of poor predictability—the difficulty is compounded. In such circumstances the design and management of a research program for acquiring desired knowledge is every bit as important as the funding itself.

The purpose of the program we recommend is not simply a general acquisition of knowledge for its own sake, nor is it a "let's *do* something" focusing of research energies on whatever aspects of the stratospheric modification topic currently seem accessible to study. The prime need is rather to obtain as early as possible the basis for rational decisions regarding desirability and details of CFM regulation.

The effectiveness of a research program depends on the quality of the investigators working in it and on the overall pattern of projects in which they are supported. Both aspects benefit from broadly based, continued

21

oversight. Therefore, as an early step in initiating such a program, an advisory council of cognizant individuals should be assembled from the academic and federal research community to recommend research directions and to review program operations. Provision should be made for renewal of this advisory council by systematic rotation of membership throughout the life of the program.

The council should form task forces to identify the pertinent, specific research questions whose answers would fill the troublesome gaps in knowledge and to plan tractable research efforts in some reasonable detail. Only when specific, feasible research directives have been established should the funding agencies solicit appropriate research proposals. The proposals received should be evaluated by appropriate task forces under the advisory council, and a continuing system of advisory peer review should be applied to all research undertaken in the program.

Rather than conduct this research through only one or two federal agencies, the government should involve several agencies, along with the academic community, in the project. Not only would this minimize any real or apparent conflicts of interest between regulatory and research functions, but it would also maximize the opportunity to involve ongoing programs already engaged in biological and climatic work related to the questions posed and would thus capitalize on the experience and resources already invested. At the same time, the advisory peer review should be able to assure that the research in the program is conducted by persons with particular competence and continuing performance.

While the main functions of such a review system would be to ensure well thought-out work by competent investigators using appropriate techniques, we believe that such a mechanism would also greatly increase the chance that really novel ideas of uncertain but possibly high promise would be funded. The council and its task forces should organize and arrange so that this will be the case.

Whatever the level of support decided for this program, it should be sustained with reasonable assurance, over a period of years, both to allow the time for recruiting the best people into it and to allow them, once recruited, the opportunity to develop some significant body of research findings. No amount of hasty expenditure, with uncertain continuity, can effectively foster the systematic growth of knowledge required.

STRATOSPHERIC PROCESSES AND CLIMATE

Although crash programs are often necessary to deal with the particular and often urgent stratospheric environmental problems (e.g., SST's, Space

Shuttle, fluorocarbons), the Committee recommends the establishment and support of a long-range program designed to improve our general understanding of the stratosphere and of the effects of stratospheric changes on climatic changes. The knowledge gained from such a program might enable the early detection of a potential problem and would aid in the more efficient management, or even avoidance, of future crash programs.

In both types of program, studies must be carried out that fall into at least the following five categories:

1. Atmospheric measurements of the concentrations of minor chemical species (and of solar radiation), both with good space, time, and altitude coverage and with maximum coincident measurement of different quantities. This will require use of both local and remote sensors and a variety of platforms.

2. Laboratory research on the physical and chemical properties of known atmospheric species and on the rates of reaction of those species with one another and with light. This includes the accurate determination of absorption cross sections, spectroscopic properties, and photolysis products over a wide wavelength range. It also includes the measurement of rate constants of chemical reactions among atmospheric species, possible pollutant molecules, and some excited-state molecules, as functions of temperature.

3. Theoretical work and model development of varying rigor and sophistication, taking account of the results of 2 above and validating its results by comparison with measured quantitites in 1 above. Atmospheric models of two quite different kinds are needed—transport models and climatic-change models. The simpler transport models rely on our knowledge of transport properties of the present climate and may combine these with detailed chemistry. So far, models with detailed chemistry have involved highly simplified, one-dimensional transport calculations. Intensive research is needed on higher-dimensional models, with more realistic transport and adequately detailed chemistry. Modeling climate change, both natural and human-produced, is much more difficult, and considerable work is already under way in this area. Research is needed with special emphasis on the climatic effects of stratospheric changes and changes in radiative properties of the troposphere produced by certain pollutants.

4. Development and general acceptance of standard methods for routine mapping and summarization of the measurements of the distribution in space and time of ozone and other low-concentration constituents. The reduction of such measurements is far from trivial and requires well-chosen allowances for fluctuations both in smoothing and in stating the uncertainties of the results.

5. Processes and mechanisms determining the natural variations of ozone, particularly the longer-term changes. As we improve our understanding of natural ozone changes, we will be able much more easily to interpret observed changes, to identify human-produced changes, and to give guidance to the assessment of historical changes in uv-B intensity.

The constraints of short-range programs are usually so severe that none of the five research categories can be advanced within them in an orderly, substantive way. Detailed recommendations for both the short- and long-range programs cannot be set down here, but their general philosophy is clear: there is a profound need for long-range, well-planned research in all five categories. The need is probably greatest in items 1 and 3, both of which are expensive and time-consuming. Vigorous, expanded laboratory work must go along with field work and modeling programs. We wish to emphasize, however, that a successful long-range research program must not be hampered by overly detailed long-term planning. It must remain flexible enough to respond to the stimuli of its own results, thus optimizing their benefits.

For a partial listing of recommended studies for the short-range program, principally dealing with the fluorocarbon problem, the reader is referred to Appendix F of the report of the Panel on Atmospheric Chemistry, *Halocarbons: Effects on Stratospheric Ozone.*

BIOLOGY

Future and continuing decisions about reductions in CFM releases will depend in part on the impacts of ozone reduction acting through the effects of enhanced solar uv-B radiation.

An important perspective on biological effects is provided by the correlation between skin-cancer incidence and uv-B exposure, as discussed under human health effects elsewhere in this report. However, a more formidable problem is understanding the effects of increased uv exposure on plants, animals, and microorganisms in both agricultural and natural ecosystems. Since the report of the Climatic Impact Committee (1975), little advance has been made in the understanding of uv effects that would contribute to the quantitative evaluation of these biological hazards.

A commitment to a sustained program of biological research is necessary, if our uncertainties about the potential effects discussed elsewhere in this report are to be replaced by reasonably confident predictions. The sought-for predictability is particularly important in light of the continuing assaults on the stratospheric ozone by diverse sources.

We therefore recommend a comprehensive research program to accomplish the following objectives:

1. Determine the major effects of solar uv-B radiation on intact plants and animals, together with how the effects are caused and the natural defense mechanisms protecting against them. This work, including the measurement of relevant action spectra, will provide information basic to accomplishment of the following objectives.

2. Establish the solar uv-B sensitivity of the major agricultural plants and animals, to permit quantitative prediction of any reductions (or increases) in economic yield accompanying likely levels of ozone depletion.

3. Develop criteria to identify those species of plants and animals in nonagricultural ecosystems that are potentially most sensitive to solar uv radiation. These criteria would permit estimation of natural ecosystem effects, without the necessity for individual investigation of the prohibitively large number of species involved.

4. Assess the perturbations in ecosystems that would be expected if the uv-sensitive species were subjected to projected increases in solar uv-B radiation.

5. Maintain, and improve the methods for, monitoring environmental uv-B radiation on a continuing, systematic basis, in order to provide a record of the solar radiation stress acting on biological systems at selected locations in the world for the immediately ensuing decades.

6. Investigate the penetration of solar uv-B radiation into aquatic ecosystems (both freshwater and salt water) and its relation to the spatial distributions of plankton in natural waters and the uv-B stress they experience. This investigation should provide a basis for making predictions of aquatic ecosystem perturbations by uv-B radiation.

7. Undertake carefully planned studies of changes in natural ecosystems associated with the 20 percent variation of uv-B radiation over a decade-long period. These studies would have to extend over at least half a cycle (6 years) and might well require as many as two full cycles (20-odd years) in order to eliminate the effects of other kinds of short-term climate variation.

The above research program should be designed to answer the questions posed as quickly as possible, recognizing, however, that this will require a sustained effort over more than 5 years and the involvement of many of the biologists already most skilled in the fields involved. An intensive "crash" program of limited duration cannot supply the needed information.

HEALTH RESEARCH

Both need and opportunity exist for a considerable variety of inquiries into the epidemiology and etiology of skin types and skin cancers. This section lists and discusses briefly some of the most obvious instances.

SKIN TYPES

Information on the national distribution of skin types (as described in Chapter 8) would be of considerable service in assessing the detailed magnitude of the U.S. risk. Information on the distribution of skin types in cities/metropolitan areas for which skin-cancer data are best would be of great service in understanding and interpreting these data.

Melanoma Epidemiology As much information as feasible on the distribution of skin types among melanoma cases should be collected, thus providing the second essential in relating risk to skin type. The interpretation of the data on detailed location of melanomas now being collected would be enhanced by information as to what fraction of the melanoma cases in a definable geographic area are being seen.

The possible usefulness of skin-typing information for relatives, over and above the information provided by the skin type of the person concerned, should be investigated, since the identification of smaller groups with higher risks would contribute much to the effectiveness of preventive measures. Local or regional variation of incidence at specific ages, particularly as reflected in the age distribution of cases, deserves careful study, since it can illuminate the roles of other independent or contributing causes of melanoma. The relation between melanoma site and occupation deserves considerable attention, as does that between site and age. The possible relation of a variety of cellular enzyme activities—enhanced, normal, or reduced—to melanoma incidence should be investigated.

Cohort studies of melanoma incidence should be carried out; they could lead to better estimates of future melanoma rates (given undisturbed damaging uv, or DUV, dose and human behavior).

Nonmelanomas Information on skin types for adequate samples of cases in selected areas would again be important in assessing relative risks. (Prospective data collection may be the only possibility.)

Modification The possible reversal of certain potentially precancerous changes in epithelial skin cells by 5-fluorouracil suggests consideration of trials, on an adequately controlled basis, of regular doses of this compound on patients reappearing for at least the third time with nonmelanomas.

MECHANISMS

A deeper understanding of the mechanisms by which uv-B exposure leads to skin cancers of various sorts would be of great value, particularly as this understanding illuminates differences in susceptibility.

PREVENTIVE MEDICINE

The near-future need is for a study, by a diverse group, of the costs and returns to be expected from an active public health program. This would include consideration of what could be taught in the schools, how effective sun screens can be most effectively made available to the public (for example, by control or special name, thus distinguishing them adequately or by incorporation in cosmetics), what costs would be involved in large-scale skin typing, and how this might be reduced.

REFERENCE

Climatic Impact Committee. 1975. *Environmental Impact of Stratospheric Flight: Biological and Climatic Effects of Aircraft Emissions in the Stratosphere.* National Academy of Sciences, Washington, D.C.

Phenomena in
the Stratosphere

While considerably more detail is needed to get trustworthy numbers, some general pictures can help us to understand the problems of the ozone shield and the chemicals—natural and man-made—that attack it.

THE ATMOSPHERE

About 87 percent (by weight) of the earth's atmosphere makes up the troposphere, which extends up about 16 km (10 miles) over the equator and about 8 km (5 miles) over the poles. Compared with the bulk of the rest of the atmosphere, it is rapidly mixed. In less than a year the troposphere in either all the northern hemisphere or all the southern hemisphere will mix quite completely. A little longer, a year or two, is required for relatively complete mixing across the equator. Since CFMs are released at the surface, whatever amount is released will soon be mixed throughout the troposphere.

Most of the rest of earth's atmospheric mass is in the stratosphere. In it, horizontal mixing is moderately rapid, but vertical mixing is slowed down by the stability associated with temperatures that are first nearly constant and then increase as we go higher and higher. Air in the stratosphere does return to the troposphere, but only, on the average, after a few years.

The troposphere contains about seven times as much air as the stratosphere, so the time taken, on the average, for air in the troposphere to return to the stratosphere is about seven times as long—two or three decades.

The boundary between the stratosphere and troposphere is called the tropopause. It is defined as the region in which there is a temperature minimum (more precisely, a sharp change in the rate of change of temperature with height). This property causes mixing between the stratosphere and troposphere to be slower than the mixing within either of those regions.

ULTRAVIOLET LIGHT

Ultraviolet light is invisible, and each of its units (photons) is more energetic than the photons of visible light. Starting near the edge of the visible, we can usefully define the following wavelength ranges (where the visible extends between 400 and 750 nm):

320 to 400 nm—uv-A

290 to 320 nm—uv-B, largely responsible for sunburn, skin cancer, and other biological effects

240 to 290 nm—does not reach the earth's surface, mainly because of the shielding effect of ozone (O_3)

180 to 240 nm—absorbed by molecular oxygen (mainly with formation of ozone) or by ozone and does not reach the lower stratosphere

below 180 nm—does not reach even the high stratosphere in appreciable quantities

The shorter the wavelength of uv-B light, the greater its biological effects. Relative effectiveness of one quantum for altering DNA, the material carrying the genetic information of all living cells, is roughly as follows:

320 nm	0.03
315 nm	0.1
310 nm	0.6
305 nm	2.6
300 nm	15
295 nm	60
290 nm	160

(These define the *action spectrum* of uv-B light for DNA, which is here normalized to a value of 1000 at 265 nm, the wavelength for peak sensitivity. At still shorter wavelengths the sensitivity drops off.)

We shall refer to the total amount of DNA-altering light, weighed by its relative effectiveness, as the damaging uv, or DUV, intensity.

Either normal or reduced amounts of ozone in the stratosphere absorb rapidly increasing amounts of light as we pass from 320 to 290 nm, where absorption is almost complete. As a result of the combination of the action spectrum and the ozone absorption, wavelengths near the middle of the uv-B contribute most to DUV intensity and to biological effects.

A reduction in the total ozone from any cause allows more ultraviolet light to reach the earth's surface. The corresponding increase in DUV is even greater than the increase in total uv-B, because the increase is proportionally greater at shorter wavelengths.

ABSORPTION OF ULTRAVIOLET LIGHT IN THE ATMOSPHERE

Absorption of ultraviolet light by molecular oxygen (O_2) (below 240 nm) is the mechanism by which O_3 is formed.

Absorption of ultraviolet light by ozone, mainly between 200 and 320 nm, leads to the conversion of light energy into heat but not to an appreciable change in the amount of ozone. This is because most of the oxygen atoms (O) formed react rapidly with oxygen molecules (O_2) to reform O_3. The result is a steady state in which ultraviolet light from the sun maintains an oxygen atom concentration proportional to both the ozone concentration and the uv intensity.

In the range (200 to 240 nm) where both molecular oxygen and ozone absorb ultraviolet light, they compete for the radiation available. Thus, the presence of more ozone will decrease the amount of light absorbed by molecular oxygen and thereby decrease its own rate of production. This process tends to stabilize, to a limited degree, the amount of ozone present.

The fraction by volume of the atmosphere made up of ozone (the mixing ratio) varies from 12 ppm (parts per million) in the stratosphere to a small fraction of ppm in the troposphere (away from urban pollution). There is no isolated "ozone layer." We will, accordingly, speak of "the ozone shield," which consists of all atmospheric ozone wherever located, or of stratospheric ozone, which is that portion (almost all) affected by the CFMs.

Photochemical smog (produced from hydrocarbons, oxides of nitrogen, and uv-A under conditions of urban pollution) contributes daytime ozone in the polluted atmosphere near the surface. In sufficiently high concentrations, this surface ozone (together with other highly reactive pollutants) is health damaging. Its additional uv-shielding contributions to the ozone shield are, indeed, negligible in comparison with the accompanying health-damaging effects.

REACTIVE COMPOUNDS IN THE ATMOSPHERE

We have already mentioned the uv-driven exchange between O_3 and atomic oxygen (O). Since both species involve an odd number (3 or 1) of oxygen atoms, and since both are highly reactive, it was natural to refer to them together as "odd oxygen."

We will extend this language to other classes of more or less reactive species whose existence and concentrations are linked together by chemical reactions of appreciable speed (but often very much slower than other fast reactions such as those that link either O_3 or O). In particular, we will speak of the following:

"odd nitrogen": mainly the compounds NO, NO_2, N_2O_5, and HNO_3.

"odd hydrogen": mainly the free radicals HO and HO_2, the compound H_2O_2, and H atoms.

"odd chlorine": mainly Cl atoms, ClO radicals, and HCl and $ClONO_2$ molecules.

Note that "oddness" in this sense does not necessarily imply an odd number of the specified atoms, since H_2O_2 is odd hydrogen and N_2O_5 is odd nitrogen.

Not all of any type of "oddness" is immediately available to attack ozone. Thus, ClO radicals and Cl atoms are active odd chlorine, while HCl and $ClONO_2$ contain passive odd chlorine. The larger fraction of the time any kind of oddness is passive, the less effective a given total amount of that kind of oddness is in destroying ozone.

In all these cases, once "odd something" is formed, the result usually circulates among the corresponding species many times before the chain is broken. Chemical reactions producing such exchanges between parts of odd species can eliminate one odd oxygen in each exchange (two eliminations per cycle of exchange). As a typical example, the basic "odd chlorine" cycle consists of the chemical reaction

$$Cl + O_3 \rightarrow ClO + O_2$$

coupled with

$$Cl + O \rightarrow Cl + O_2$$

which, together, transform "odd oxygen," O and O_3, into two oxygen molecules. The great potency of "odd" nitrogen, hydrogen, and chlorine depends on the many times such a catalytic cycle can occur once "oddness" of hydrogen, chlorine, or nitrogen has been created.

The formation of chlorine nitrate, in which both odd chlorine and odd nitrogen are passive, thus decreases both the amount of active odd chlorine and the amount of active odd nitrogen that would otherwise be present. Given the relatively large amount of odd nitrogen available at lower stratospheric altitudes, adding odd chlorine can produce an increase in ozone at those altitudes. This is because, in effect, more odd nitrogen becomes passive (as chlorine nitrate) than new active odd chlorine is produced. At higher altitudes, quite the reverse happens, mainly because lower pressures reduce the formation of chlorine nitrate. Thus, the balance can be, and is, quite different at different altitudes. Further allowances has to be made for the increased amount of quite short-wavelength ultraviolet light, responsible for decomposing chlorine nitrate, that is present at the higher altitudes.

THE FATE OF N_2O

One of the most important processes of plant life is "the nitrogen cycle" in which nitrogen from the atmosphere is first incorporated in the plant and later, often after the plant's death, returned to the atmosphere. This return, largely of N_2 itself, is in part in the form of other nitrogen-containing gases, such as NO_2, NO, and N_2O. All of these, except N_2O, are either water-soluble or easily converted into water-soluble compounds and are not expected to escape from the troposphere before being "rained out." Whether land plants or (microscopic) ocean plants are the major source of natural N_2O is still a matter of debate.

The N_2O moves into the stratosphere, again rather slowly. At heights above 25 to 30 km, some of it is attacked by the most active form of oxygen $O(^1D)$ and is converted to NO and NO_2, thus providing the reactive nitrogen species that play the dominant catalytic role in the destruction of ozone. Both in the lower stratosphere and in the troposphere, these gases are in equilibrium with HNO_3, which is subject to being "rained out." They are thus removed from the atmosphere once they come down into the troposphere.

DESTRUCTION OF OZONE (MORE GENERALLY, THE DESTRUCTION OF ODD OXYGEN)

The amount of ozone in our ozone shield depends basically on how fast it is created (absorption of far uv light by O_2) and on how fast it is destroyed. The main destruction reactions involve:

— reaction with atomic oxygen (normally about 20 percent)
— reaction with "odd nitrogen" (normally about 60 percent)
— reaction with "odd hydrogen" (normally about 10 percent)
— reaction with "odd chlorine" (probably about a few percent)

The best available model calculations now suggest that a continuing release of CFMs at the 1973 rate, worldwide, would, after a few decades

—decrease ozone at the higher levels (say at 40 km) perhaps by as much as 25 percent.

—decrease ozone by much smaller amounts at lower altitudes

—on balance decrease the total ozone shield by about 7 percent

NOTICE CAREFULLY that all these figures would be changed, some considerably, if various reaction constants were changed to other values that are also consistent with the present extent of our knowledge. We give single specific numbers here to paint a general picture, and to stress the difference in effect at different altitudes, NOT to give answers to be accepted without allowance for the many uncertainties.

The main points to be drawn from this illustrative example are these:

—The average height at which ozone absorbs its share of solar radiation will be reduced.

—Reductions at higher altitudes are considerably larger than the average reduction and are thus particularly sensitive to the values of reaction rates, etc.

THE FATE OF F-11 AND F-12

What happens when either of the two chlorofluoromethanes of greatest concern is released into the atmosphere?

First, relatively rapid tropospheric mixing occurs. Then slower "leakage" into the stratosphere and more or less random ascent to a height where the appropriate kind of ultraviolet light (185 to 225 nm) will be encountered. This usually takes place at heights above 25 km (16 miles). This entire process requires several decades for the average molecule and thus introduces long delays in some of consequences of changes in amounts released.

The CFM now absorbs such ultraviolet light and forms "odd chlorine," initially one odd chlorine for one CFM molecule but eventually one for each chlorine in the CFM molecule. A small amount of chlorine will leave the stratosphere almost immediately. On the average, however, it will remain in the stratosphere for a few years. Once it reaches the troposphere, it will be "rained out" mainly as HCl.

In the course of its residence, each odd chlorine will destroy an average of thousands of ozone molecules. The number destroyed will depend upon:

— the rates of various chemical reactions

— the rate of return to the troposphere

Thus, we can appreciably enhance the destructivity of the stratosphere for ozone by the infiltration of a relatively small amount of F-11 or F-12.

Carbon tetrachloride (CCl$_4$), probably largely man-made, is today a serious source of stratospheric chlorine—perhaps comparable with F-11 and F-12—but its rates of release are decreasing and only a fraction of those of F-11 and F-12.

Other chlorine-containing compounds, such as F-22, F-21, or methyl chloride, also enhance the destructivity of the statosphere for ozone. They do this, however, to a much more limited extent, because they are attacked and, in a large measure, decomposed in the troposphere, where the odd chlorine produced will be rained out almost at once.

SCENARIOS

The picture just given helps us to understand the general behavior over time of the increase in DUV intensity consequent on a given schedule of CFM release. If CFMs were to be released at a constant rate, the concentration in the troposphere would build up toward a steady-state value. To come close to this steady-state value would take a few times as long as the average time for tropospheric air to pass through the stratosphere—several decades.

The CFM mixing ratio in the stratosphere is lower than that in the troposphere because

1. (vertical) diffusion and circulation processes in the stratosphere are not instantaneous,
2. photolysis destroys the CFMs in the high stratosphere.

A sudden cessation of CFM release, if attainable, would "freeze" the tropospheric concentration of the CFMs. Since this tropospheric concentration would, in such a scenario, have been increasing, the concentrations at various levels in the stratosphere would be lower than corresponds to a steady state with this tropospheric concentration. Thus, it would take several years for the stratospheric concentrations of the CFMs to rise to the maximum (a small multiple of the few years involved in returning stratospheric air to the troposphere). Since in this scenario the total amount of CFM in the atmosphere can only decrease, the stratospheric concentrations would then decrease by a substantial factor over several decades (a small multiple of the 2 to 3 decades needed, on the average, for tropospheric air to pass through the stratosphere). Thus, on this scenario of sudden and complete cessation, the ozone reduction would increase for several (actually about 10) years after cessation—and would take many decades to decrease to a small fraction of its value at the time of cessation.

Balancing
Losses and Gains

NATURE OF THE CONTROL PROBLEM

If control of substances, like CFMs, which have many uses, is needed, what should be the character of the control? To some, it is natural that there be a total allowable release, to be rationed out among the various uses. To others, it is natural that each use be controlled separately, balancing its benefits against the disadvantages that it causes. The choice between these positions ought not be a matter of philosophy. Rather it ought to depend on how the disadvantages increase as the release schedule is increased. For simplicity, let us think of doubling the whole of an entire release schedule, throughout its history. Would the (unfavorable) consequences double? Or would the disadvantages more than double were the release schedule doubled? Or would they be less than double? (We will use doubling as an example to make the discussion easier to follow, recognizing that the real choices would probably involve much smaller factors as final regulation came into balance.)

If, as one possibility, successive equal increases in the release schedule had seriously increasing effects, it would be both appropriate and important to set a total bearable release and then allocate this total to the most important uses.

If, as another possibility, release schedules differing only by a constant factor had disadvantages roughly proportional to their size, so that adding a particular whole schedule had effects at least roughly independent of background, it would be appropriate to regulate each use wholly separately, balancing the advantages of use against the disadvantages of release.

If, as a third possibility, a given release of CFMs were somewhat less important against a larger background, so that doubling the release schedule produced somewhat less than a doubling of disadvantages:
— We could reasonably begin by regulating each use separately, on a basis of balancing disadvantage and advantage (since we would be close to the second possibility).
— As the accumulation of CFMs in the atmosphere was decreased, we would expect to find it desirable to be somewhat more stringent in our regulation. (In view of the long times involved in passage to the stratosphere, this would take decades, not years.)

Similarly, if a given release were somewhat more important against a larger background, so that doubling the release schedule produced more than a doubling of disadvantages:
— We could reasonably begin by regulating each use separately, on a basis of balancing disadvantage and advantage.
— As the accumulation of CFMs in the atmosphere was decreased, we would expect to find it desirable to be somewhat less stringent in our regulation. (In view of the long times involved in passage to the stratosphere, this would take decades, not years.)

Which of these four situations actually applies for CFMs? The first rough approach to an answer is that, since we are only contemplating small decreases in the ozone shield, effect will be nearly proportional to cause at each step, so that, overall, the disadvantages of release will be nearly proportional to the amount released. As we shall see, this rough conclusion is nearly correct—to be sure of this, however, we need to examine the relationships involved a little more closely.

Consider two release schedules, in one of which, year after year, the rate of CFM release is double the other (both corresponding to small percentages of reduction in the total ozone shield). The rates of CFM decomposition by ultraviolet light will also be a two-to-one ratio, as will the total amount of odd chlorine. What about the ozone depletion?

As noted above, a reduction of, say, 5 percent in the total ozone involves a larger percentage reduction of ozone at higher altitudes, say 40 km, and a smaller percentage reduction at lower altitudes, say 25 km. As a consequence, doubling the odd chlorine at all altitudes less than doubles the effect at higher altitudes. The consequence is that the effect on the ozone shield of doubling the release schedule would be somewhat less than a doubling of the ozone depletion.

For small ozone depletions, uv-B increases are nearly proportional to ozone depletions. As percentage depletions become larger, the process accelerates; but in the range of a few percent of depletion, behavior is quite close to proportionality. As a result, doubling the release schedule would produce rather less than a doubling of the excess uv-B arriving at the ground. The relation of skin cancers to uv-B dose is not one of simple proportionality [e.g., see Climatic Impact Committee (1975), page 42]. Rather the logarithm of the incidence seems proportional to the dose. This means some acceleration. Doubling the excess uv-B arriving at the ground will rather more than double the excess load of skin cancer. A more detailed examination shows that this acceleration is, however, small for small percentages of increase. (This acceleration thus compensates—possibly over-compensating somewhat—for the failure of the uv-B dose to grow proportionately to the increase in the schedule of releases.)

As a consequence, in the one case where we understand in numerical detail how the release of CFMs produces a disadvantageous effect, a doubling in the release schedule is to be expected to produce roughly a doubling of the number of additional skin cancers. Thus, regulation use by use, comparing the advantages of each use with the disadvantages of the resulting release, appears to be the appropriate form of regulation.

A similar conclusion, less well founded, plausibly applies to climate effects.

GRADED RESPONSE

Given that the benefits and costs of the various uses of F-11 and F-12 deserve to be balanced separately, what are the consequences if and when control of CFM use becomes appropriate?

In general, the less essential the use, the more rapidly and vigorously we ought to be prepared to curb it. At one extreme stands the use of chlorofluoromethanes as spray-can propellants, where a variety of alternatives are available. Particularly since this use has accounted for more than one half of all releases, there would be good reasons to curb this use rapidly and significantly, if control of CFM releases becomes needed.

At the other extreme, the use of F-11 and F-12 for household refrigeration of food is very important for human health. Even with roughly 70 million refrigerators and 20 million freezers in service in the United States, less than 1 percent of CFM emissions from all uses results from this use. Clearly curbing use of CFMs in household refrigeration is not likely to be appropriate.

There clearly could be no reason for curbing spray-can uses and household refrigeration uses with the same vigor and over the same time scale. Spray cans currently provide more than 200 times the amount of release that household refrigeration does, and in most cases have served a relatively minor and replaceable use.

Other uses of CFMs fall in between these extremes. Of them, automotive air conditioners are an important example. Their usefulness is greater than that of spray cans and less than that of household refrigerators. They are estimated to provide more than one third of all CFM emissions from refrigerant uses, much more than any other refrigerant use. Of all emissions from mobile air conditioners, about 90 percent has been estimated to be eliminatable by design changes, by changes in service practices, or by recovery of residual charge when equipped cars are junked. Here regulation, if and when required, should clearly be at an intermediate vigor and speed, focusing first on service and junking practices, for which immediate results are possible, and then on redesign. The development of satisfactory mobile air conditioners not using F-11 and F-12 should be encouraged, and their successful development should be followed with a phase out of production of units using F-11 and F-12.

More information on the relative importance of releases associated with these and other uses will be found in Appendix D.

REFERENCE

Climatic Impact Committee. 1975. *Environmental Impact of Stratospheric Flight: Biological and Climatic Effects of Aircraft Emissions in the Stratosphere,* National Academy of Sciences, Washington, D.C.

Expected Changes
in Ozone — Findings

The following is the summary chapter of the report of the Panel on Atmospheric Chemistry, *Halocarbons: Effects on Stratospheric Ozone.* Their report has been most carefully prepared; much hard work has gone into not only best estimates but also carefully assessed ranges of uncertainty.

PANEL ON ATMOSPHERIC CHEMISTRY

H. S. Gutowsky, *Chairman*
University of Illinois at Urbana-Champaign

Julius Chang
Lawrence Livermore Laboratory

George Pimentel
University of California at Berkeley

Robert Dickinson
National Center for Atmospheric Research

H. I. Schiff
York University

Dieter Ehhalt
Institute for Chemistry, West Germany

John H. Seinfeld
California Institute of Technology

James P. Friend
Drexel University

Brian Thrush
University of Cambridge

Frederick Kaufman
University of Pittsburgh

Cheves Walling
University of Utah

R. A. Marcus
University of Illinois at Urbana-Champaign

Geoffrey Watson
Princeton University

Bruce N. Gregory, *Executive Secretary*
Richard Milstein, *Staff Officer*

39

INTRODUCTION

The stratosphere is a region that extends from about 16 to 50 kilometers (52,000 to 160,000 feet) above the surface of the earth at low latitudes and from 8 to 50 kilometers at high latitudes. In contrast to the lower atmosphere, where there is turbulence and vertical mixing, the stratosphere is relatively quiescent. As a consequence, it is particularly susceptible to contamination, because pollutants introduced there tend to remain for long periods of time—several years or more.

One of the trace constituents of the stratosphere is ozone (O_3). Although ozone represents only a few parts per million of the gases in the stratosphere, potential threats to this ozone have become a focus of scientific interest and public concern during the past few years. This is because, even in its small amount, stratospheric ozone absorbs virtually all of the solar ultraviolet (uv) radiation with wavelengths less than 290 nanometers (nm) and most of that in the biologically harmful 290 to 320 nm (uv-B) wavelength region, thus preventing the radiation from reaching the surface of the earth, where it could adversely affect human, plant, and animal life. As a consequence of this absorption, and that in the visible and infrared, ozone also helps to maintain the heat balance of the globe and is directly responsible for the temperature inversion (temperature increasing with increasing altitude) that characterizes the upper stratosphere. Thus, any decrease in ozone would increase the amount of harmful uv-B radiation reaching the earth's surface; it would also perturb the atmospheric heat balance and thereby might trigger a change in the world's climate.

Concern over human effects on stratospheric ozone was first raised as a possible consequence of emissions from high-flying aircraft. This particular problem was intensively studied by scientists during the Department of Transportation's Climatic Impact Assessment Program (CIAP). The effects of aircrafts and other activities of man upon stratospheric ozone depend on the natural processes that determine the distribution of ozone in the stratosphere, unperturbed by man. Understanding of those natural processes is extensive and on a demonstrably firm foundation. The ozone distribution is maintained as the result of a dynamic balance between creation and destruction mechanisms. Ozone is produced in the upper stratosphere by the action of solar uv radiation upon molecular oxygen and is destroyed by several processes. The most important of these, which accounts for about two thirds of the total destruction rate, is a catalytic chain reaction involving various oxides of nitrogen (NO_x). Other relevant destruction mechanisms include direct reaction of oxygen atoms with ozone (Chapman reaction) and catalytic chain reactions involving several species containing hydrogen or chlorine (HO_x and ClO_x).

The stratospheric production of ozone is relatively insensitive to man's activities. The rate is determined by the intensity of solar radiation of wavelengths shorter than 242 nm, as well as by the distribution in altitude of molecular oxygen and of the ozone itself. The absorption of solar radiation by pollutants can affect the amount and distribution of the uv light that is available to dissociate oxygen. In this indirect way, pollutants can affect the ozone production, but such secondary effects are small.

The time required to destroy an ozone molecule can, however, be influenced appreciably by man's activities. As mentioned above, several naturally occurring catalytic chemical reactions have been identified as ozone-destruction mechanisms. The chemical species involved in these reactions (NO_x, HO_x, and ClO_x) are referred to as catalysts because they are not used up by the reactions. The individual reactants are regenerated and thereby are capable of reacting with ozone over and over again. Each of them can remove thousands of ozone molecules, before being destroyed itself by some other process. Consequently, even though the concentration of these catalytic molecules in the stratosphere is quite low (1 to 10 parts in 10^9), they have important effects.

Unfortunately, artificial introduction of these catalysts into the atmosphere in the large amounts now associated with man's activities can lead to a significant increase in their stratospheric concentrations. As a consequence, the average lifetime of an ozone molecule is decreased relative to that in the unperturbed stratosphere. Since the overall production of ozone is not increased, while the individual molecules are destroyed more rapidly, the result is a net reduction in the amount of ozone present. One such example of human ability to modify stratospheric ozone is the direct emission of NO_x into the stratosphere from the exhausts of SST's and other high-flying aircraft, referred to above. Another is the release of chlorofluoromethanes (CFMs) in the use of spray cans, air conditioners, and refrigerators. To give an idea of the magnitudes involved, 1 percent of the global ozone is about 33 million metric tons; the total world production of the CFMs $CFCl_3$ (F-11) and CF_2Cl_2 (F-12) in 1974 was nearly a million metric tons.

The main purpose of the present report is to evaluate the extent to which stratospheric ozone will be affected by the CFMs and other chlorine compounds introduced by man. We also consider the hydrogen chloride and particulates emitted by the Space Shuttle that is now being developed. Moreover, in the course of our work there have been suggestions that additional stratospheric pollutants, such as nitrous oxide (N_2O), derived from nitrogen fertilizers, or methyl bromide (CH_3Br), used as a fumigant, could produce appreciable reductions in stratospheric ozone. Our treatment of them has been limited to assuring that their effects are less immediate than those of the CFMs. Nonetheless, the importance of stratospheric ozone to life on earth requires that all such suggestions be thoroughly investigated and that attention be focused upon the aggregate effects.

THE CFMS CFCl₃ (F-11) AND CF₂Cl₂ (F-12)

The two CFMs most widely used and about which there has been the greatest concern are $CFCl_3$ (F-11) and CF_2Cl_2 (F-12). *All the evidence that we examined indicates that the long-term release of F-11 and F-12 at present rates will cause an appreciable reduction in the amount of stratospheric ozone. In more specific terms, it appears that their continued release at the 1973 production rates would cause the ozone to decrease steadily until a probable reduction of about 6 to 7.5 percent is reached, with an uncertainty range of at least 2 to 20 percent, using what are believed to be roughly 95 percent confidence limits. The time required for the reduction to attain half of this steady-state value (3 to 3.75 percent) would be 40 to 50 years.* *

There is little question about the fundamental aspects of the problem. F-11 and F-12 have been produced and used in large quantities that are a matter of quite accurate record. The large fraction of the production that enters the atmosphere can be readily inferred from the types of use. The compounds do accumulate in the troposphere; they have been measured there at steadily increasing concentrations consistent with the estimated release rates. Moreover, recent measurements confirm that F-11 and F-12 are transported into the stratosphere. Laboratory experiments show that once in the stratosphere F-11 and F-12 must undergo photolytic dissociation to produce Cl atoms. The CFM concentrations have been observed to decrease in the middle and upper regions of the stratosphere at a rate corresponding to the combined effects of photolysis and the "transport lag." Finally, as soon as the Cl atoms are generated in the stratosphere, they will react with O_3 in the catalytic cycle by which Cl and ClO destroy ozone. These reactions have been measured individually in the laboratory, and they must occur in the stratosphere. It is inevitable that CFMs released to the atmosphere do destroy stratospheric ozone. The more difficult problem is evaluating such effects quantitatively.

The numerical values for ozone reduction by F-11, F-12, and other pollutants are determined in general by the aspects just described—the amounts released, transport in the atmosphere, and the particular photochemical and chemical reactions involved. There is, however, the important qualification that alternative removal mechanisms for these pollutants, if any exist, could

---

*For a discussion of uncertainties in the time dependence see Chapters 5 and 8 of the Panel report.

reduce the results accordingly. Moreover, there might be other processes tending to diminish or to amplify the effects upon stratospheric ozone of the ClO_x generated from the CFMs. As in the case of all physical and chemical phenomena, none of these factors can be measured exactly. There are uncertainties in each, and much of our effort has been spent on identifying the possibilities and uncertainties and reducing them to the extent feasible within the time available. Each of these aspects is reviewed separately below.

Many of the uncertainties in predicting the ozone reduction are difficult to evaluate. Whenever possible we have given numerical estimates, as ranges or percentages ($\pm$) about the value that seems most probable at this time.* Frequently, for the normal, symmetrical distribution of measured values, the experimental errors are expressed in terms of the standard deviation about the average value ($\pm\sigma$). We have elected to use uncertainty limits *equivalent* to two standard deviations ($\pm2\sigma$). This, of course, does not affect the uncertainties, but it may influence the way in which they are perceived. If one takes the smaller $\pm\sigma$ range, the probability that the actual value falls between $+\sigma$ and $-\sigma$ is 68 percent, i.e., these are 68 percent confidence limits. There is one chance in three (32 percent) that the actual value lies outside the $\pm\sigma$ range. Because of the importance of the ozone reduction, it seems better to use wider limits that are more likely to include the actual value. With uncertainty ranges corresponding to $\pm2\sigma$, the confidence limits are 95 percent; there is only one chance in twenty that the actual value falls outside that (larger) range.

RELEASE RATES

The recent surveys of F-11 and F-12 production, use, and release rates cited in Chapter 3 (of the Panel report) have improved significantly the completeness and reliability of the data available. The totals for the amounts produced and released so far are more accurate than the annual figures, which require year-end inventory estimates. The data from Eastern Bloc countries are still approximate, but these involve a minor part (5 percent) of the total. *The stated uncertainty in the total amounts of F-11 and F-12 that have been released so far (through 1975) has now been reduced to ±5 percent.*

The calculations of ozone reduction are for specified release rates, so errors in the actual release rates do not affect directly the numerical predictions. However, the uncertainties in release rates do enter when one applies

---

*The shapes of the probability distributions are also uncertain in most of the cases considered; the contributing factors are quite different, of doubtful symmetry and conceivably "long-tailed."

the predictions to the actual releases. Also, the uncertainties in release rates are highly important in determining whether there are unknown processes removing CFMs from the atmosphere.

TRANSPORT

The history of a particular pollutant molecule from point of release to the time of its degradation and/or removal from the atmosphere as long as 50 or 100 years later on the average, is complicated by the atmospheric motion. Furthermore, once ClO_x is generated from a CFM, one needs to know what happens to it. The reduction in stratospheric ozone is the net effect of enormous numbers of such histories. The calculation and adding up of the histories to obtain the net effect is a time-consuming mathematical problem made feasible only by replacing particular histories with averages. Some averages involve time period (diurnal and seasonal); others are for location (latitude and longitude). The type of averaging employed defines a model for the atmospheric motions and chemistry and introduces characteristic approximations (see Chapter 7 of the Panel report). These approximations produce uncertainties in the results in addition to those associated with the various constants that describe the rates of reaction.

So far, the calculations available to us of the effects of the CFMs on strato-spheric ozone have been made by the one-dimensional (1-D) model. The approach employed is equivalent to averaging the concentrations, motions, and reactions over latitude and longitude, leaving only their dependences on altitude and time. There are physical reasons why such simplification is reasonable; for example, any longitudinal (east–west) differences are expected to be small. In any case, the equivalents of averaged transport rates appear in the model as the vertical eddy-mixing or transport coefficient K, which depends only on the altitude. One approximation made in the 1-D model is the choice of K, which is adjusted empirically to fit the experimentally observed distribution in altitude of trace substances in the atmosphere.

A closely related approximation, less apparent but implicit in the 1-D model, is the use of space- and time-averaged concentrations to calculate the reaction rates that determine the ozone reduction. The nature of the approximation is described in Chapter 7 of the Panel report, but its accuracy is difficult to establish. These approximations can be avoided, in principle, with two- or three-dimensional (2- or 3-D) calculations. However, this was not feasible within the time period established for our study. In practice, as described in Chapters 5 and 7 and Appendix B of the Panel report, K was chosen to fit the altitude profile (concentration as a function of altitude) observed for a tracer gas, such as N_2O or CH_4, that is itself released at ground level and undergoes destruction in the stratosphere. The effects of the chemistry are included in the 1-D calculations made to develop the fit;

hence the choice of K depends on the correctness of the tracer-gas chemistry involved as well as on the concentration profiles used.

We have explored the approximations of the 1-D model in three ways. First, a thorough investigation was made of the procedures used to choose K (Chapter 5 and Appendix B of the Panel report) and how these affect the uncertainties of CFM histories in the atmosphere. Second, concentration profiles were calculated with various choices of K for a number of important, reactive atmospheric species and compared with the all-too-limited measurements that have been made so far (see Chapter 7 of the Panel report). Also, some comparisons are made with the results of 2-D calculations. Finally, as described in Chapter 8 of the Panel report, ozone reductions were calculated using different choices of K but keeping other model parameters the same. Much of the interpretation of these diverse studies is subjective, and combination of their results in an overall numerical uncertainty range is very difficult. However, the various comparisons generally agree well within a twofold or at most a threefold range.

At present, we estimate that use of the 1-D model to approximate the distribution and transport of the chemical species involved in the reduction of stratospheric ozone by the CFMs causes uncertainty by a factor of 1.7 in either direction (+70 to −40 percent) in the predicted amount of the globally averaged reduction (a threefold range).

STRATOSPHERIC CHEMISTRY

In addition to the transport approximations just described, each calculation is subject to uncertainties in the quantitative details of stratospheric chemistry. Some workers may treat the chemistry more completely than others or use more efficient computer programs, but, in principle, any 1-D calculations can include all the known photochemical and chemical processes affecting ozone. The difficulty of concern here is that the factors governing the catalytic destruction of ozone by CFMs for any *specific* reaction scheme are subject to experimental error in their determination. These factors include the chemical reaction rate constants, the solar flux, the photolysis rates, the temperature distribution in the stratosphere, and the concentrations (or source and sink strengths) of trace species in the unperturbed atmosphere (see Chapter 7 and Appendix D of the Panel report). Of these, the largest source of uncertainty that we have identified is a relatively small number of reactions that dominate the ClO_x chemistry in the stratosphere. These reactions involve unstable, highly reactive species, and the determinations of some of their rate constants are extremely difficult. Lesser uncertainties are identified with the photochemical processes and with the concentrations of natural species (Chapter 8 of the Panel report).

The sensitivity of the calculated ozone reduction to the rate constants, or to any other input parameter, can be investigated simply by calculating the reduction for different values of the particular input parameter, keeping all other aspects of the 1-D model the same. We have done this with the rate constants for seven reactions for which the uncertainties presently have a large effect on the outcome, obtaining the results given in Table 8.3 of the Panel report. *For the particular reaction scheme employed, uncertainties in seven of the rate constants cause a fivefold uncertainty range in predictions of ozone reduction by the CFMs. The largest contributions are from the HO + HO₂ and HO + HCl reactions; the other reactions included in the analysis are ClO + NO₂, Cl + CH₄, ClO + O, ClO + NO, and Cl + O₃. Additional uncertainties in the photochemical processes and the concentrations of natural species are estimated to increase the overall uncertainty range associated with the stratospheric chemistry to a factor of 2.5 in either direction (a sixfold range).*

The initial calculations of ozone reduction by the CFMs were more sensitive to experimental errors in the rate constants than was *generally* appreciated at the time, because of the lack of systematic studies such as those described in Chapter 8 of the Panel report. Moreover, the recent inclusion of $ClONO_2$ in the reaction scheme has increased the contribution of these rate constants to the uncertainty of the predictions from a fourfold to a fivefold range. The sensitivity studies show that the dependences of the ozone reduction on changes in individual rate constants are largely independent of one another, within the ranges given for each. Therefore, the consequences of future improvement in the seven rate constants can be estimated by a simple scaling of the results given in Table 8.3 of the Panel report.

OTHER FACTORS

So far, our discussion assumes that all the CFM released will contribute to the ozone reduction according to the reaction system employed in the calculations (Appendix D of the Panel report). However, if this is not the case, and some additional mechanism modifies the effects of the CFMs, the ozone reductions otherwise calculated would have to be scaled up or down accordingly. Several types of possibilities have been considered: inactive removal (that does not lead to ozone destruction), competing reactions, feedback mechanisms, and large natural sources of stratospheric chlorine.

Inactive Removal The importance of inactive removal may be seen by considering a particular pollutant such as F-12 (CF_2Cl_2) under steady-state conditions, for a given release rate. Photolysis and reaction with $O(^1D)$ in the stratosphere remove about 1 percent per year of the total amount of F–12 in the atmosphere, giving products that destroy ozone. If they are the only removal processes, the atmospheric residence (or removal) time of F–12 is nearly 100 years. If, however, there were also a process (sink) that removed an additional 2 percent per year, but did not destroy ozone, the fraction of F-12 that destroyed ozone would be one third instead of unity, and the ozone reduction would also be one third of what it would otherwise be. Similarly, the total residence time and the amount of F-12 in the atmosphere would be multiplied by 1/3 in the steady state for this case, which would be attained more rapidly.

We see that inactive removal processes become significant when their rate approaches or exceeds the overall rate of removal via stratospheric photolysis. That removal, however, is a slow process, and therefore other processes need not be very fast to compete with photolysis and be important. In principle, since any natural sources of F-11 and F-12 are negligible, the best way of searching for such processes would be to carry out an *accurate* materials balance, i.e., a comparison of what has been released into the atmosphere with what is measured to be actually present plus what is calculated to have been used up in the stratospheric processes that destroy ozone. Any missing CFM would be the amount removed without affecting stratospheric ozone. However, such comparisons are of limited value unless the total amount in the atmosphere (the global burden) and the amount released are *both* known to high accuracy (≤ 5 percent).

Attempts have been made at a materials balance, as presented in Chapters 3 and 6 and Appendix E of the Panel report. More extensive atmospheric measurements are available for F-11, which is easier to detect, than for F-12, which is more abundant. Unfortunately, even for F-11 the observations are limited. The inadequacy of available observations combines with the difficulty of the measurements to give an uncertainty in the global atmospheric burden that is too high (± 40 percent) for the comparison with the release rates that are known with greater accuracy (± 5 percent) to have much significance. Although the materials balance can be interpreted as consistent with little or no inactive removal of the CFMs, the uncertainty limits range from zero inactive removal to a rate sixfold faster than that of the stratospheric photolysis (see Appendix E of the Panel report).

We have also taken the other approach to this question—looking at each of the individual inactive removal mechanisms. An intrinsic difficulty with this approach is that an important possibility might be overlooked. The large number of suggested mechanisms may be classified according to the site of removal (surface, troposphere, or stratosphere) and according to the nature of the process, e.g., a nondestructive reservoir or chemical degradation. Examples of the latter two categories that have been frequently cited are incorporation in the polar ice caps and decomposition via reactions with neutral species (HO, O, etc.) in the troposphere. Neither is significant; the rate of the first is demonstrated to be no more than 0.001 percent per year, while for the latter the concentrations involved and/or the reaction rates are small. In Chapter 4 and Appendix A of the Panel report these and the other possibilities suggested have been analyzed carefully on the basis of known chemical reactions and known physical processes. The results are summarized in Table 4.2 of the Panel report.

Three processes have estimated inactive removal times for F-11 and F-12 that are short enough to warrant further, more detailed study. Lower limits of ~ 10^2 (70 and 200), 10^3, and 5 × 10^3 years have been placed, respectively, on the removal times for solution in the surface waters of the oceans (followed by some unknown degradation process) and by ion-molecule reactions and photodissociation in the troposphere. If each of these processes actually removed F-11 and F-12 in the time corresponding to the lower limit set for it, the maximum combined effect would be a decrease in the predicted ozone reductions by at most 2/5 of what they would be in the absence of such inactive removal. However, we expect the effect to be no more than 20 percent (a decrease by 1/5), based on the limited data available for the oceanic sink.

Competing Reactions Stratospheric processes that remove Cl or ClO from the ClO_x catalytic chain limit the amount of ozone that they destroy. Thus, the formation of HCl and $ClONO_2$ (by the reactions Cl + CH_4 and ClO + NO_2+M) provides temporary "reservoirs" that store the chlorine from decomposed CFMs in inactive forms pending downward transport from the stratosphere followed by rain-out from the top of the troposphere. There are also reactions that convert HCl and $ClONO_2$ back into active species, so that the importance of the reservoirs depends on the balance struck between formation and reconversion.

At first, HCl was considered to be the only reservoir of any consequence for the ClO_x catalysts. However, a re-examination in early 1976 of the possible role of $ClONO_2$ indicated that $ClONO_2$ might be a significant reservoir, doubly important because it removes not only ClO from the ClO_x cycle but also NO_2 from the NO_x cycle. Since then, the intensive laboratory studies of the formation and destruction processes for $ClONO_2$ (see Appendix A of the Panel report) have confirmed its probable importance in stratospheric chemistry. Therefore, its reactions have been incorporated in our calculations and in the results we report (Appendix D and Chapter 8 of the Panel report). Its inclusion reduces the predicted ozone reductions by a factor of about 1.85, modifies to some degree the distribution with altitude of the ozone, and increases the kinetics-related range of uncertainty in the ozone reduction from fourfold to fivefold.

The important role of HCl is supported by stratospheric measurements of its occurrence, as compared with a calculated distribution in Chapter 7 of the Panel report. In the case of $ClONO_2$, infrared measurements have placed an *upper bound* of ~1 ppb on its current stratospheric concentration at 25 km (Chapter 6 of the Panel report). The corresponding concentrations calculated for that region are on the order of 0.5 ppb (Figure 7.16 of the Panel report). Therefore, more sensitive measurements are needed to provide a definitive check on the inclusion of $ClONO_2$ in the reaction scheme; such observations should be available in the latter part of 1976.

Another possible competitor with the catalytic effectiveness of ClO is its photolysis, but this proves too slow to be important (Appendix A, Section III of the Panel report). A wide variety of other processes that might affect Cl and ClO have been considered. One example is the coupling between the ClO_x and NO_x cycles (see Chapter 9 of the Panel report), which occurs in addition to that provided by $ClONO_2$. It is included in the reaction scheme employed (Appendix D of the Panel report), along with a number of less important processes. The probable importance of $ClONO_2$ was a surprise. Whether or not there is another such surprise in store remains to be seen. However, a modest number of reactants is involved, the number of their possible reactions is finite, and these possibilities have for the most part already been examined.

Feedback Mechanisms   Several types of feedback mechanisms have been proposed that might alleviate or amplify the effect of CFMs on stratospheric ozone. One is the partial "self-healing" of the stratosphere. This argument states that if ozone is destroyed in the upper stratosphere by CFMs (or other pollutants), the uv solar radiation penetrates deeper into the stratosphere, photolyzing more O_2 and generating more O_3 at lower levels. This does occur to some extent, but its importance is limited, and the effect is included in the 1-D calculations that we have employed (see Chapter 7 of the Panel report).

Another, more speculative suggestion is that redistribution of the ozone to lower altitudes by the effects of the CFMs will increase the temperature at the tropical tropopause and in the lower stratosphere and enable more H_2O to enter the stratosphere. This would, in turn, generate more HO radicals by reaction of H_2O with $O(^1D)$, to convert HCl back to catalytically active Cl by the important HO + HCl reaction, and increase the rate of ozone destruction. A better understanding of water vapor transport between troposphere and stratosphere and of the temperature changes is needed before this mechanism can be regarded as established. But it might have a significant effect, and its further study is desirable. Details are given in Chapter 9 of the Panel report, along with comments about other less likely possibilities.

Natural Sources of Stratospheric Chlorine It has been suggested that the injection into the atmosphere from natural sources of large amounts of chlorine compounds, such as HCl, CH_3Cl, and perhaps CCl_4, casts doubt upon or reduces the significance of the man-made sources. Such suggestions have not borne up under close scrutiny. Although the "natural" chlorine was not included in the early calculations, this has since been done, and, as described in Chapter 7 of the Panel report, the effects are modest. The importance of HCl is reduced because of its rapid "washout" from the troposphere by rain, and that of CH_3Cl by destruction processes in the troposphere. The present total reduction in stratospheric ozone by HCl, CH_3Cl, and CCl_4 (from whatever sources) is calculated to be less than 1 percent. These three compounds now contribute *roughly* the same amount of chlorine to the stratosphere as do the CFMs (Chapter 3 and 6 of the Panel report), and the ozone reductions they produce are also comparable.

The most important fact, however, is that the reduction in stratospheric ozone by chlorine from man-made sources is increasing and will be in addition to whatever is caused by chlorine from natural sources (Chapter 3 of the Panel report). The latter are already at their steady-state amounts, while continued release of CFMs at recent rates will probably cause their atmospheric concentrations, and ozone reduction, to increase tenfold or more. The significance of the man-made sources would be reduced if the effects of ClO_x were nonlinear in the correct direction, i.e., if the stratospheric addition of chlorine from CFMs to that from natural sources produced a less than proportionate increase in the ozone reduction. Such effects are to be expected only at catalyst concentrations that give reductions of ozone greater than 15 to 20 percent, compared with an ozone reduction of ≤ 1 percent by ClO_x from natural sources.

We have examined a variety of proposals that might alleviate or amplify the reduction of stratospheric ozone by CFMs, such as competing reactions, feedback mechanisms, and natural sources of chlorine. Several of them have been incorporated in our calculations; others have been eliminated as inconsequential; a few are considered unlikely to have major effects but warrant further attention. The role of $ClONO_2$ as an inert reservoir for stratospheric chlorine seems particularly important, the available data indicating that it reduces the effects of the CFMs by a factor of nearly 2 (1.85).

Predicted Ozone Reduction and Its Overall Uncertainty With inclusion of $ClONO_2$ in the reaction scheme, the reduction in stratospheric ozone by the CFMs is predicted to be 7.5 percent at steady state for constant 1973 release rates (Chapter 8 of the Panel report). The analyses summarized above indicate that this value might be decreased by about 20 percent if inactive removal by an oceanic sink does indeed occur. If this is confirmed by future measurements, it could reduce the predicted value of the ozone reduction to 6 percent.

The approximations and uncertainties involved in predicting this value have been reviewed above. Each source of uncertainty produces a distribution of less likely values for the ozone reduction about the central value of 6 to 7.5 percent. Thus, insofar as the atmospheric chemistry taken alone is concerned, its sixfold range of uncertainty means that the real value for the reduction might be as little as 2.4 percent or as large as 15 percent, i.e., 6 to 7.5 percent times (1/2.5) or 2.5, with roughly a 95 percent chance of finding the real value between these limits. Equivalent statements apply to the other two sources that we have estimated in numerical terms. The three distributions are combined to give a new distribution of uncertainty. Their uncertainties are independent so we express them as multiplicative factors (times f) and combine their effects.* The possibility that some as yet unidentified processes might affect the predictions is *not* included in the analysis.

The combination of the multiplicative uncertainty factors for the release rates (1.05), transport (1.7), and stratospheric chemistry (2.5) leads to an overall eightfold uncertainty range. Application of these limits to the central values (6 to 7.5 percent) of the reduction in ozone expected after many decades of releasing F-11 and F-12 at the 1973 rates gives a range of about 2 to 20 percent, for the uncertainties from these three sources and the oceanic sink.

Insofar as we know, this is the first detailed attempt to assess the overall uncertainty in the calculation of the reduction in stratospheric ozone by the CFMs. The tenfold range reflects the limitations of our knowledge as well as our use of what we believe to be demanding (95 percent) confidence limits. The fivefold range for rate constants reduces the relative effects of any other, smaller uncertainties, provided they are not systematic in one direction or the other. In considering the implications of these results, it is important to remember that the most probable value is about 6 to 7.5 percent, that there is roughly a 95 percent chance of the real value being between the 2 to 20 percent limits given, and that while one might prefer a particular limit, both limits must be considered.

VERIFICATION OF PREDICTED OZONE REDUCTIONS

Ideally, one would like to have a direct check on the ozone reduction predicted for the CFMs or at least some more direct means than are now available for narrowing the uncertainty of the predictions. However, either objective will take time to accomplish, as well as effort. If the release of CFM continues during that time, it will increase the eventual ozone

---

*The square root of the sum of $(\ln f)^2$ was calculated for the three sources to obtain r. The values of $\exp(-r)$ and $\exp(+r)$ are the factors that give the lower and upper limits for the ozone reduction, when multiplied by the most probable value (6 to 7.5 percent).

reduction beyond what it would have been if CFM release had been curtailed. The amount of such increase would depend on the differences between the two CFM release schedules compared, the period of time involved, and the actual extent of ozone reduction per unit of CFM (Chapter 8 of the Panel report).

Direct observation of a decrease in stratospheric ozone attributable to the CFMs is obscured by the natural, long-term irregularities of about ± 5 percent that are still incompletely understood (Chapters 9 and 6 of the Panel report). Furthermore, the long-term trends must be determined in the presence of much larger daily, seasonal, and latitudinal changes, with possibly a small component associated with the sun-spot cycle thrown in for good measure. Even if a change occurs in the long-term trend, there is the problem of deciding whether it is natural or due to the CFMs. This requires a large enough reduction by the CFMs over a long enough period of time to identify its growth characteristics (as inferred from model calculations) on top of the natural background. Sophisticated statistical analyses are being applied to the problem, and better observations for the purpose are being gathered.

So far, the ozone data have provided no case for the global stratospheric ozone having been decreased (or increased) by the CFM releases. Nor would one expect a ~ 0.5 percent reduction, that estimated to have already been produced by past CFM releases, to be detectable compared with the natural fluctuations of ± 5 percent. Moreover, the detection and identification of an ozone reduction as small as 2 percent by the CFMs would require carefully calibrated ozone observations extending over several years, either by an improved network of surface stations or balloons or by satellite, together with adequate statistical analysis of the data. Some scientists are more optimistic than others about the length of time and magnitude of reduction required for success.

A more immediate approach is to measure the concentrations of the other key reactants that establish the reduction in ozone. In fact, most of the reactants have been demonstrated to be in the stratosphere at the levels corresponding to those assumed in, or predicted by, the calculations of the reduction in stratospheric ozone. For example, the stratospheric profiles observed for total NO and NO_2 agree with the calculated profiles to within 50 percent (Appendix C of the Panel report). However, observations in the stratosphere of several of the most important but highly reactive species of low concentration are either very few (O, HO, ClO) or still to be accomplished (HO_2, Cl). Although the catalytic cycle employed in the analysis undoubtedly exists, detailed and careful measurements of these species would help to reduce the possibilities of unknown factors that might affect the extent of ozone reduction actually produced by the cycle. In particular, stratospheric measurements of Cl and ClO, because of their direct removal of O_3 and O, should be especially valuable in attributing an actual decrease in ozone to the CFMs, after allowance for Cl and ClO from other sources.

Certainly, much has been learned about the CFM problem during the past two years. In fact, our evaluation of it has been a case of "shooting at a rapidly moving target." Further improvements will occur during the next year or two. These will include a more complete understanding of atmospheric chemistry, better determinations of the rate constants and absorption coefficients, and improved atmospheric measurements, most desirably on a global scale. The limits on inactive removal processes should be more closely defined; more direct evidence of the amount of reduction in ozone by the CFMs should be provided by observations of Cl and ClO in the stratosphere; and the approximations made in the predictions should be improved by the application of 2- and 3-D models to the problem.

Continuation of CFM releases at their present (static) levels, while waiting for improvements in our ability to determine the ozone reduction caused by the CFMs, will increase the eventual peak ozone reduction and its total (integrated) amount that actually occurs in comparison with what they would be if release were curtailed at once. Each year of release will increase the peak reduction by about 0.07 percent (central value of a 0.02 to 0.2 percent range) and the total amount of reduction (integrated over time) by 1/10. A resumption of exponential growth would of course give annual increments of increasing size.

EFFECTS OF OTHER POLLUTANTS

Besides the CFMs, there are several other pollutants that require mention. These are reviewed below to give a catalog of the growing number of ways in which man may reduce stratospheric ozone and to consider some of the implications.

THE SPACE SHUTTLE

As now planned, combustion of the solid propellant in the Space Shuttle will inject HCl gas and aluminum oxide particulates directly into the stratosphere. However, the amount of chlorine introduced into the stratosphere per year by 50 flights per year (the number now projected for 1986) will be only about 1 percent of that from the CFMs for continued release at the 1973 rates; and the effects will be relatively small (Chapter 9 of the Panel report). Similarly, the amount of particulates is modest compared with that naturally present from volcanic action; and, as described in Chapter 9 of the Panel report, there is no reason to believe that such materials have significant effects on stratospheric chemistry. *We conclude that the combustion products from the Space Shuttle at the presently planned launch schedule of 50 per year will make a small contribution (~0.15 percent with a range of 0.05 to 0.45 percent) to the total reduction of stratospheric ozone by human activities.* Further-more, since these products are injected directly into the stratosphere, their atmospheric residence time is relatively short, so there would not be long-lasting aftereffects should the program be terminated.

N_2O FROM FERTILIZER

The natural abundance of stratospheric ozone is determined to a large degree by the NO_x produced from N_2O. In turn, nitrogen fertilizers contri-bute to the amount of N_2O released. These facts, in combination with the increasingly widespread use of fertilizers, have led to a number of studies, now in progress, of the possible future impact of nitrogen fertilizers upon stratospheric ozone. The data presently available are inadequate to judge the issue (Chapter 9 of the Panel report). *More detailed studies of the reduc-tion in stratospheric ozone associated with the use of nitrogen fertilizers are essen-tial, especially of the biological production of N_2O and the mechanisms for its removal from the troposphere.*

OTHERS

The report of the Climatic Impact Committee dealt with the NO_x emissions from large fleets of several hundred SST and high-altitude subsonic planes, projected for 1990. These were estimated to reduce the stratospheric ozone by significant amounts in the absence of adequate emission controls. (Climatic Impact Committee, 1975, p. 29, Table 4; Grobecker et al., 1974). Other possible ways in which man may contribute to the reduction in strato-spheric ozone include F-22 ($CHClF_2$) used largely for refrigeration, CH_3Br employed in agricultural fumigation, and a number of hydrogen-containing and unsaturated chlorocarbons listed in Table 3.2 of the Panel report. At present release rates, the ozone reduction that will be caused by each is rela-tively minor (≤ 0.05 percent); however, the use and release of some of these substances seem likely to increase. Furthermore, although it is not yet clear whether the CCl_4 now found in the atmosphere is largely man-made or natural in origin, it has an appreciable effect—about 0.5 percent.

THE TOTAL BURDEN OF POLLUTANTS

The reductions in stratospheric ozone by ClO_x from different sources are additive to a useful approximation for total reductions of no more than 15 to 20 percent. Beyond this point, nonlinear responses will begin to be important. Similar considerations apply to NO_x catalysis. However, there are interactions among different reactive species (e.g., ClO_x and NO_x) from different pollutants that cause their effects on stratospheric ozone to depend on the amount of other species present. The formation of $ClONO_2$ from ClO and NO_2 is one example, and the importance of the amount of water present is another (Chapter 9 of the Panel report). Also, it is likely that nonlinear responses will occur, at some value(s) of the ozone reduction for the biological, climatic, and any other consequences. Hence, we must not only be concerned about the sum of the individual effects, i.e., the total burden of pollutants placed upon the stratosphere, but also be alert for interactions among them. *We find that procedures must be established to follow the production and release of pollutants that affect stratospheric ozone, to monitor their concentrations in the atmosphere, and to analyze their combined effects.*

SUMMING UP

In judging the consequences of the findings it seems important to bear in mind several features of the analysis that have not yet been emphasized. For purposes of simplicity, the disucssion so far has been mainly in terms of the ozone reduction calculated for steady-state conditions (at the 1973 release rates for F-11 and F-12). This is the most widely used of the various possibilities for presenting the predictions. However, the time scale of events is highly significant. This may be seen in the ozone reduction calculated for a constant CFM release rate (1973) until 1978, when all release is halted (Figure 8.5 of the Panel report). The ozone reduction continues to grow for a decade beyond cutoff (or cutback) and then requires an *additional* 65 years to recover one half of its maximum loss. A competitive sink of 1 percent per year would change but little the maximum ozone reduction of this scenario, but it would accelerate the subsequent recovery.

Another important aspect of the findings is that the ozone reduction calculated for the CFMs lies in a critical range of values. If it were an order of magnitude smaller, it might be viewed as relatively minor. If it were an order of magnitude larger, a reduction probably would have been detected by now and action taken to curtail release of the CFMs. Therefore, one must consider the likelihood and consequences of future increases in release rates of the CFMs.

When the CFM problem was first recognized two years ago, CFM use had had two decades of exponential growth at a rate of 10 percent per year. If such growth had continued, the uncertainties we have discussed translate into uncertainties in the time required to achieve the release rate corresponding to a particular ozone reduction at steady state. However, the actual releases in 1975 and 1976 experienced a 15 percent drop from the exponential growth curve and are comparable with the 1973 release rate used by us for the steady-state predictions. But there is no assurance that future releases will remain constant. *Resumption of exponential growth in the production and use of* CFMs *could well occur and lead to a doubling of their release rate within 10 years. Even if the release rates became constant at that point, they would cause a doubling in the expected ozone reduction, to a value of about 12 to 15 percent, with a range of 4 to above 25 percent, once a steady state was reached.* .

Furthermore, CFM production and use are worldwide. If U.S. release is curtailed but other use continues the more rapid exponential growth evident in Figures 3.1 and 3.2 of the Panel report, the magnitude of the overall reduction in stratospheric ozone could still reach much higher levels even though a longer time would be required. *Clearly, although any action taken by the United States to regulate the production and use of* CFMs *would have a proportionate effect on the reduction in stratospheric ozone, such action must become worldwide to be effective in the long run.*

Finally, while our knowledge of stratospheric ozone has become extensive during the past few years, it should be apparent from the discussion given above that significant uncertainties remain. An interim report of the Panel was issued in July 1975, identifying a wide range of observations, experiments, and actions needed to deal more adequately with the CFM problem in particular and with threats in general to stratospheric ozone. Many of these needs are brought into sharper focus by the problems faced in preparing the present report, as summarized in Appendix F of the Panel report. *Additional improvements in our knowledge of the atmosphere and of stratospheric chemistry are essential to permit more accurate assessments to be made of the extent of potential reductions in the stratospheric ozone.*

REFERENCES

Climatic Impact Committee. 1975. *Environmental Impact of Stratospheric Flight: Biological and Climatic Effects of Aircraft Emissions in the Stratosphere.* National Academy of Sciences, Washington, D.C.

Grobecker, A. J., S. C. Coroniti, and R. H. Cannon, Jr. 1974. *Report of Findings: The Effects of Stratospheric Pollution by Aircraft.* Department of Transportation, Washington, D.C.

*****

COMMENT BY THE COMMITTEE ON
IMPACTS OF STRATOSPHERIC CHANGE

The use of the term "confidence limits" in this chapter needs some careful consideration. In situations where numerical uncertainties are faithfully reflected in differences between, for instance, duplicate measurements, statisticians have detailed and careful techniques for setting confidence limits, which can then be given specific probability interpretations. In most cases of measuring physical constants, such as the speed of light or the reaction rate constants involved here, any attempt to specify the uncertainties adequately will depend on skilled judgment and will produce larger uncertainties than those assessable from the apparent repeatability of the results, whether obtained in one laboratory or in more than one. Judgment is even more deeply involved in such matters as assessing eddy-mixing profiles, also very important here. The "95 percent confidence limits" quoted above are based on the best judgments the Panel was able to make. Thus, they can be given no specific probability interpretation and do not call for the refinements of calculation appropriate to the simpler situation, where measured differences do faithfully respond to all sources of error or variability. They are the best facsimiles the Panel knows how to provide of what correct "95 percent confidence limits" would be. They deserve interpretation, in the Panel's judgment, in the same way that more formally obtained "95 percent confidence limits" would be interpreted wherever their calculation was appropriate.

Impacts of Increased Cfms
on Climate

GENERAL CONSIDERATIONS

The presence of CFMs in the atmosphere, even in very small concentrations, may have significant effects on climate. Such effects may be direct or indirect.

Direct climatic effects may follow from the fact that CFMs absorb and emit radiation in the infrared "window" region of the electromagnetic spectrum, where the earth's atmosphere is otherwise relatively transparent. Increased concentrations of CFMs may thus trap thermal (infrared) radiation emitted from the surface and warm the lower layers of the atmosphere.

Indirect climatic effects arise from the possible destruction and redistribution of ozone by the odd-chlorine products that result from the photolytic decomposition of CFMs. When ozone is removed, more ultraviolet and visible radiation reaches the ground, which tends to warm the lower atmosphere and the earth's surface. At the same time, the loss of ultraviolet absorption by ozone in the stratosphere reduces heating there. Consequently, less thermal (infrared) radiation is emitted toward the ground (particularly in the 9.6-μm absorption band of ozone). This tends to cool the lower atmosphere and the earth's surface. The two effects compete against one another to result in a reduced net influence on the climate of the lower atmosphere. Because the changes depend not only on the total ozone but also on the redistributions of ozone in both altitude and latitude, quantitative evaluations of the effects are, as yet, insufficiently precise to establish

58

Natural climatic variability complicates the assessment of the growing human impacts on climate, such as those that may occur following CFM releases to the atmosphere. In statistical analyses aimed at isolating the human impacts by empirical inference, the natural variations of climate typically result in very low signal-to-noise ratios and therefore in a low level of confidence in the results of the analyses. Only if the human impacts were very large, and increasing rapidly with time, would they be clearly recognizable above the noise level, in a reasonably short interval of years. Impacts that are relatively small at present (but that may become large in the future and must therefore be carefully assessed) would be likely to escape notice altogether.

In theoretical analyses aimed at assessing human impacts by physical inference, many of the complexities of climatic behavior, which may affect the ultimate character of the impacts, are not yet able to be incorporated into such analyses in a completely realistic manner. To the limited extent that attempts have been made to take these complexities into account, however, the fact that the consequences of each are difficult to isolate from one another (because of highly nonlinear interaction effects) introduces uncertainty that—rather like the climatic "noise" encountered in statistical analyses—tends to lower the confidence that we may have in the results of the analyses.

In the general context of assessing human impacts on climate, the most promising avenue to understanding appears to be through numerical modeling of the "climatic system" (Figure 1). As progress is inevitably made toward the more realistic simulation of the behavior of each part of the climatic system and of the interactions between the different parts of the system, progressively more reliable and more detailed estimates of the human impacts can be expected in future years.

In a limited way, numerical climate models are already able to contribute to the goal of assessing human impacts through so-called sensitivity studies. In such studies, comparative calculations are made both with and without the change of an externally specified parameter of the model, capable of simulating the introduction of a particular human disturbance into the climatic system.

Most sensitivity studies to date are based on constant-climate (equilibrium-climate) models that do not include the mechanisms responsible for the characteristic variability of climate. We may find that models that adequately predict the natural variations of climate will yield a response to a perturbation that depends on the prevailing variation. Should this be true, i.e., should there be interaction of serious magnitude, we would then need to assess the effects of man-made changes against the background pattern of natural climatic fluctuations to be experienced specifically during the next 50 years. Information on simply the mean effects, averaged over a

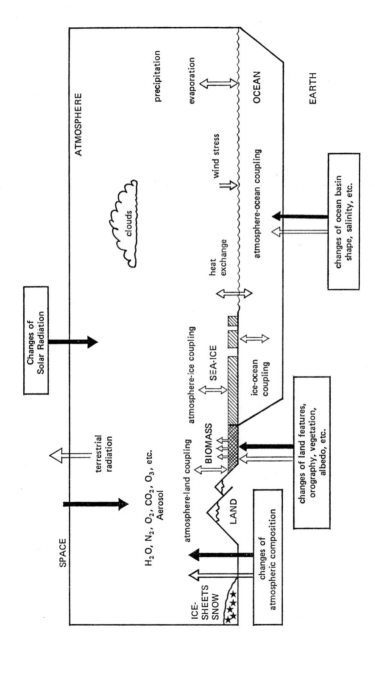

FIGURE 1 Schematic illustration of the components of the coupled atmosphere-ocean-ice-land surface-biota climatic system. The solid arrows are examples of external mechanisms, and the open arrows are examples of internal mechanisms in climatic change.

wide variety of such patterns, might not be sufficient. Thus, we attach special importance to efforts to develop variable-climate models that have broad predictive capabilities, rather than the relatively narrow, specialized capabilities that characterize most present-day climate models.

Although a confident assessment of CFM impacts on climate remains beyond our grasp today, it is instructive to summarize the contribution of such impacts—as best we understand them—in the context of the contributions of other sources of climatic variability, both natural and human-generated. An attempt to provide this kind of perspective is shown in Table 4, in which the factors identified as numbers 6 and 7 are those specifically related to CFM effects. If nothing else, the table illustrates the difficulty of attributing observed climatic variations to any single causal factor and the necessity of investigating the problem of CFM impacts on climate in a physical framework sufficiently general to encompass a wide range of interacting influences.

CLIMATIC EFFECT OF INFRARED ABSORPTION BY CFMS

It has been shown by Ramanathan (1975, and a more detailed preliminary note, unpublished) that the presence of F-11 and F-12 in the atmosphere with tropospheric concentrations of 2 ppb each would be sufficient to decrease the outgoing flux of global terrestrial radiation, everything else remaining constant, by about 0.5 percent. For comparison, concentrations of 0.7 ppb and 1.9 ppb are those that would be asymptotically reached at steady state, by F-11 and F-12, respectively, if future releases of these CFMs were to continue indefinitely at their 1973 rates. (Eighty percent of these asymptotic concentrations would be reached by the year 2050 in the case of F-11, and by the year 2125 in the case of F-12.) These calculations of projected concentrations are based on estimates of atmospheric removal rates as determined by the Panel on Atmospheric Chemistry (2 percent per year for F-11 and 1 percent per year for F-12). Using Ramanathan's (1975) equations, which were subsequently shown to give results virtually identical to those based on a detailed radiative–convective model (see Ramanathan, 1976), an atmosphere containing the 1973 release rate asymptotic concentrations of F-11 and F-12 (0.7 and 1.9 ppb, respectively) would result in a decrease in the outgoing flux of global terrestrial radiation of 0.3 percent and in an associated surface temperature increase of about 0.5°C.

The infrared absorption effects of CFMs may be compared with those of carbon dioxide, which in turn have been more extensively investigated on the basis of several different models. Especially noteworthy among the

TABLE 4 Possible Causal Factors in Future Climatic Change (Temperature, Rainfall, Cloudiness) to the A.D. 2000

Origin	Factor	Confidence[a] That Factor will Change Appreciably	A Confidence[a] That A Change in Factor Would Appreciably Affect Climate	Time Scale(s) of Variation to be Considered	Estimated Principal Climatic Effects[b]
Solar	1. Total solar output	Low	High	Months and longer	Warming-cooling (not clear)
	2. Ultraviolet and other variations	High	Low-moderate	Days and longer	Cooling (0.1-1°C)
Volcanic	3. Stratospheric particle injections	High	Moderate-high	Years and longer	Warming (0.1-1°C[d])
Human generated	4. Carbon dioxide increase	High	Moderate-high	Trend	Warming cooling
	5. Particle increase	Moderate	Low-moderate	Days and longer	Warming cooling
	6. Chlorofluoromethane (CFM) increase	Moderate[c]	Moderate-high	Trend[c]	Warming (0.1-1°C)[c]
	7. Ozone depletion by CFMs, NOx, etc.	Moderate[c]	Moderate	Trend[c]	uv-B radiation increase (10%)[c]
	8. Thermal pollution	High	High (local effects)	Trend	Warming, local clouds/storms
	9. Land-use changes	Moderate	Moderate (regional effects)	Decades and longer	Temperature/ precipitation changes
Oceans[e]	10. Sea-surface temperature variations	High	Moderate-high (regional effects)	Months and longer	Temperature/ precipitation changes
Cryosphere[e]	11. Seas-ice/snow-cover variations	High	Moderate (regional effects)	Months and longer	Temperature/ precipitation changes
	12. Polar ice-sheet surges	Low	High	Years and longer	Rise in sea level, possible glaciation
Biota[e]	13. Vegetation changes in response to climate	Moderate	Moderate (regional effects)	Years and longer	Temperature/ precipitation changes

[a] Confidence based on intuitive judgment of many atmospheric scientists, considering state-of-the-art knowledge.

[b] All numerical values are order-of-magnitude estimates for earth as a whole; regional effects may differ substantially.

[c] Cumulative effect by A.D. 2000 if future CFM emissions continue at the 1973 level or increase 10 percent per year.

[d] Cumulative effect by A.D. 2000.

[e] These changes are, of course, part of the overall climate process.

Adapted from: Report of North American Phase I Conference on "Living with Climatic Change," Toronto, November 1975 (Science Council of Canada, 1976).

latter are two simple radiative–convective models (cf. Manabe and Strickler, 1964; Manabe and Wetherald, 1967) and a sophisticated three-dimensional (3-D) climate model (Manabe and Wetherald, 1975). In each case the problem addressed was that of assessing the change in climate resulting from doubling of atmospheric CO_2 (in most cases from 300 to 600 ppmv). A doubling of atmospheric CO_2 is a reasonable possibility within the next 50 to 100 years. The simple models gave an equilibrium surface temperature change (which is to be interpreted as a global average) of the order of $+2°C$. The more sophisticated 3-D model (which, however, also omits the potentially important feedback produced by altered cloud distributions) suggests that with a doubling of CO_2 to 600 ppmv the global average surface temperature would increase by almost $3°C$ above its present level, and that temperatures in the polar regions would locally increase by a factor 2 to 3 times greater than the global average increase. The 3-D model also suggests that the intensity of the hydrologic cycle (both rainfall and evaporation) would increase by about 7 percent. Such changes could lead initially to a general shift of the earth's climatic belts, and ultimately to a significant melting of polar ice and a worldwide increase in sea level.

The models with which the latter calculations were made give temperature changes essentially linear in the logarithm of the CO_2 concentration (Manabe, 1976). Thus an increase in CO_2 from 300 ppm to about 350 ppm, about as large a change as has occurred since 1890, would be computed (on the 1-D model) to give a global temperature change of about $0.5°C$. Thus, if, as we expect, these models give correct comparative results, the ultimate temperature change due to CFM releases constant at the 1973 level, which would not occur for several decades, would be about the same as that resulting from the last several decades of releases of CO_2.

It should be repeated for emphasis that cloud feedback and other processes have been omitted in the models from which such conclusions are drawn. It cannot, however, be assumed that inclusion of the omitted factors would necessarily reduce the magnitude of CO_2 or CFM impacts as estimated from the simple models used here. Inclusion of such factors might actually turn out to increase the impacts. Thus actual effects may be either greater or less than the simple models predict. The various estimates of the comparative effects of CFMs and CO_2 notwithstanding, it is to be emphasized that the effects of real concentrations of CFMs are *additive* to and in the same direction as the effects of CO_2.

The key to understanding the rather extraordinary effectiveness of CFMs in altering the earth's radiation balance and equilibrium temperature lies in recognizing that CFMs have absorption bands that span about one half of the atmospheric infrared window region (at wavelengths of 8–12 mm). These

TABLE 5 Comparative Future Changes of Atmospheric CFM and CO_2 Concentrations, and of Associated Global Mean Surface Temperature, for Two Scenarios as to Future CFM Release Rates

Year	DVM Release at 1973 Rates CFM Concentration (ppb)		ΔT_s(CFM) (°C)	CFM Increasing Production and Release at 10% yr^{-1} CFM Concentration (ppb)		ΔT_s(CFM) (°C)	CO_2 Release Data and other Assumptions Based on Model of Machta(1973) CO_2 concn. (ppm)	ΔT_s (CO_2)[a] (°C)
	F-11	F-12		F-11	F-12			
1975	0.09	0.21	0.06	0.09	0.21	0.06	330	0
1980	0.15	0.29	0.09	0.17	0.32	0.09	340	0.1
1990	0.25	0.44	0.1	0.50	0.80	0.3	360	0.3
2000	0.32	0.58	0.2	1.4	2.1	0.7	390	0.5

[a] ΔT_s calculated as change due to CO_2 added since 1975. A constant of 0.4°C should be added to calculate the change due to CO_2 added since 1890.

bands have large absorption coefficients in a wavelength region that in the absence of CFMs is relatively transparent to radiation (i.e., optically thin). The CO_2 bands, on the other hand, are nearly completely absorbing at present-day CO_2 concentrations, so that the effects of increasing CO_2 are, relatively speaking, not so dramatic. This is why amounts of CFMs (2 ppb each) about one hundred thousandth of CO_2 concentrations (above 300 ppm) can have effects worthy of note.

The effects of CFMs and CO_2 on surface (global average) temperature may be compared in more specific terms for various scenarios of future CFM releases. Two such scenarios are displayed in Table 5, in which WTs is the temperature change at the surface if radiative-convective equilibrium were attained with the concentrations of substances indicated in the adjacent columns. It is evident that for continued CFM releases at the 1973 rates, the temperature effects due to the CFMs would be less than half of those due to CO_2 changes by the year 2000. It is also evident that for the scenario postulating increasing releases of CFMs, at the rate of 10 percent per year, the incremental effects of the CFMs become equal to those of CO_2 before A.D. 2000, thereafter the CFM effects actually dominate the CO_2 effects.

Further climatic effects of the CFM concentration changes may also be possible. For example, the Manabe-Wetherald three-dimensional model gives a general increase in precipitation of the order of 1 percent for such changes in radiation absorption as those considered here.

It should be stressed that such global-average changes of temperature and precipitation would be associated with much larger regional and local changes in many parts of the world. The magnitude and geographical pattern of the changes cannot yet be reliably specified in detail. The climatic impacts of both CFMs and CO_2 on agriculture, for example, would be likely to depend much more strongly on the regional and local changes than they would on the global-average changes indicated above.

The estimated climatic effects of CFM releases (see Table 5), noted here, are sufficiently large that the continued release of CFMs, at their 1973 rates, must be recognized now as potentially serious. A condition of long-continued growth of CFM releases, of as little as a few percent per annum would, to the best of our present knowledge, lead ultimately, perhaps in a century or two, to climatic change of drastic porportions.

We are drawn to the more general conclusion that the introduction into the atmosphere of long-surviving (inert) substances with strong absorption bands in the 8−12 mm wavelength and having atmospheric concentrations as low as a few parts per billion must be viewed as having potential climatic consequences. This suggests an obvious criterion for testing and screening compounds sooner or later released to the atmosphere.

REFERENCES

Joint Organizing Committee, Global Atmospheric Research Program. 1975. The Physical Basis of Climate and Climate Modelling. GARP Publications Series No. 16. International Council of Scientific Unions and World Meteorological Organization.

Machta, L. 1973. Prediction of CO_2 in the atmosphere, pp. 21-31, in *Carbon and the Biosphere, Proceedings of the 24th Brookhaven Symposium in Biology,* Upton, New York, May 1972. G. M. Woodwell and I. V. Pecan, eds. U.S. Atomic Energy Commission (Conf.-720510).

Manabe, S. 1976. Geophysical Fluid Dynamics Laboratory. Personal communication.

Manabe, S., and R. F. Strickler. 1964. Thermal equilibrium of the atmosphere with convective adjustment. *J. Atmos. Sci.* 21:361-385.

Manabe, S., and R. T. Wetherald. 1967. Thermal equilibrium of the atmosphere with a given distribution of relative humidity. *J. Atmos. Sci.* 24:241-259.

Manabe, S., and R. T. Wetherald. 1975. The effects of doubling the CO_2 concentration on the climate of a general circulation model. *J. Atmos. Sci.* 32:3-15.

Ramanathan, V. 1975. Greenhouse effect due to chlorofluorcarbons: Climatic implications, *Science* 190: 50-52.

Ramanathan, V. 1976. Radiative transfer within the earth's troposphere and stratosphere: A simplified radiativeconvective model. *J. Atmos. Sci.* 33:1330-1346.

Science Council of Canada. 1976. Living with climatic change (Proceedings, Toronto conference workshop November 17-22, 1975). Ottawa.

Nonhuman Biological Effects

GENERAL

For moderate changes in atmospheric ozone content, the change in ultraviolet (uv) flux at the surface of the earth involves only radiation on the long-wavelength edge of the Hartley and Huggins absorption bands, around 290–320 nm. The effects on plants and animals of an increased flux in this "uv-B" spectral region can presently be inferred from experiments on lower organisms, which are easy to manipulate, combined with more direct, but much less complete studies on the higher plants and animals themselves. Unfortunately, quantitative extrapolation of uv radiation effects at 254 nm—which has been used for most photobiological experiments—to effects in the 290–320-nm band is not easily accomplished (Nachtwey, 1975).

The photochemical basis of a number of biological uv effects is known. Proteins and nucleic acids, which together comprise over 90 percent of the dry weight of cells, can absorb directly at wavelengths below about 320 nm and undergo a variety of photoreactions as a consequence (Varghese, 1972; McLaren and Shugar, 1964; Kaluskar and Grossweiner, 1974). Changes in these major macromolecules can also be caused by photoreactions with natural or synthetic materials that absorb at wavelength longer than those absorbed by the biological macromolecules themselves (Lamola, 1974; Spikes, 1968). In either case, the altered molecules fail to perform their biochemical functions properly, to the detriment of the living system in which they reside.

67

For viruses and microorganisms, the uv photobiology stemming from these photoreactions is now fairly well delineated. As a result we know that damage to the nucleic acid, which carries an organism's genetic information (always DNA in the case of a cell), becomes significant at much lower levels than those at which damage to other cellular components is important. On the other hand, a number of molecular repair processes for eliminating damage to DNA exists in cells. In the normal cell these processes can erase well over 99 percent of the initially produced photodamage (see Appendix C), and cells deficient in such repair are much more uv-sensitive than their normal counterparts. A multiply mutant bacterium with complete deficiency of repair, for example, is prevented from reproducing by the occurrence of essentially a single photochemical event in its entire DNA complement (which has a total molecular weight of 3×10^9). Since this amount of damage can be created by 10 to 20 seconds of midday sun in southern U.S. latitudes, cells facing a normal daylight environment are under strong evolutionary pressure to maintain effective DNA repair processes. The existence of such repair, and its varying effectiveness under different metabolic conditions, prevents simple predictions of the effect of a given radiation exposure, even when the photochemistry and quantitative photochemical yields are known. Only the net, unrepaired damage affects a cell, and its amounts—unlike the initial damage—may not be proportional to the uv-B exposure. Furthermore, a rare type of damage resistant to repair would be more detrimental than a more common type effectively corrected.

The cellular mechanisms inhibited by photochemical alteration of DNA include two very fundamental processes: (1) DNA replication (Swenson and Setlow, 1966) and (2) transcription of the genetic information from DNA into messenger RNA for directing cellular protein synthesis (Michalke and Bremer, 1969). These processes and their regulatory mechanisms are essential to all types of living things, from the simplest to the most advanced. Without replication of its genetic material a cell cannot make another cell like itself, i.e., cannot reproduce. Without transcription it cannot utilize the genetic information of its own DNA to carry out its essential functions. Interference with either or both processes is sufficient to cause abnormal cellular behavior.

Parallel studies of uv effects in cells of successively higher organisms (including flowering plants and mammals) show that the same principal kinds of photochemical damage to DNA occur in them as are found in microorganisms, along with at least several of the same repair systems for removing damage. The implication is that these cells are harmed by uv damage to their DNA in a way similar to microorganisms and maintain similar defense mechanisms for protection. The supporting evidence is, however, less explicit, and the relative importance of direct absorption of the radiation versus absorption by sensitizing pigments in many cases is less certain. As far as known, the fundamental cellular mechanisms disrupted are the same in both higher and lower systems, but because of the cooperative behavior required of the cells that comprise a single large organism, the consequences of damage can appear in a wider range of events than simply inhibited growth and reproduction of individual cells.

The action spectra $A(\lambda)$, expressing the relative effectiveness of different wavelengths λ for causing reproductive failure of viruses and microbial cells, strongly resemble the absorption spectra of nucleic acids (Giese, 1964). This would be expected if direct absorption of the radiation by nucleic acids, rather than absorption by sensitizing pigments, is responsible for most of the inactivating photochemical damage. (Effects from radiation absorption by the aromatic amino acids of proteins in the region of 280 nm are apparent in some cases, but these spectra for the most part still resemble the absorption spectra of nucleic acids.)

The net, unrepaired, inactivating damage produced by a wavelength distribution $I(\lambda)$ of polychromatic radiation would be proportional to the integral

$$\int_{\Delta\lambda} I(\lambda) A(\lambda) d\lambda$$

which needs to be evaluated only over the region $\Delta\lambda$, where the integrand remains appreciably different from zero. For sunlight at sea level, $I(\lambda)$ is essentially zero for $\lambda < 295$ nm, while with an action spectrum corresponding to a nucleic acid absorption spectrum $A(\lambda)$ becomes very small for $\lambda > 320$ nm.

In larger organisms, $A(\lambda)$ can be modified somewhat by the filtering action (absorption and scattering) of structures overlying the cells of concern, but if DNA damage is important, the effective wavelengths would still be similar to those active on cells of lower organisms. If effects due to energy absorbed in RNA also happen to be important, the same solar wavelengths would remain responsible, since the wavelength dependence of RNA absorption is very like that of DNA. Even effects due to uv absorption in proteins should follow a roughly similar action spectrum in this wavelength region. Thus, any photochemical damage stemming from solar uv radiation absorbed directly by one of the major types of macromolecules in cells would be affected by changes in stratospheric ozone. On the other hand, effects due to absorption by sensitizing pigments could depend on wavelengths somewhat longer than those affected by ozone absorption and thus be little augmented by an increase of solar flux between 295 and 320 nm.

It is far from clear that photochemical changes in nucleic acids and proteins are the only uv radiation effects of concern, but it is certain that at least these alterations must be important, therefore, any increase in solar radiation levels between 295 and 320 nm merits close attention.

Effects of uv radiation on intact higher organisms include both acute damage to the outermost cells (e.g., sunburn in man) and chronic effects stemming from repeated or long-term exposure (e.g., skin cancers in humans and animals, diminished growth in plants). Action spectra are known for a number of acute uv radiation effects. Between 290 and 320 nm, these spectra sufficiently resemble nucleic acid absorption spectra (with or without some filtering) to make the latter a reasonable guide (Caldwell, 1971). Practical difficulties prevent the determination of similar action spectra for effects due to repeated or long-term exposure, but plausible arguments can be made that they are similar to action spectra for acute effects. These arguments can be supported in some cases by further evidence (see, for example, Setlow, 1974). Therefore, the amount of biological damage from solar uv radiation under any prescribed solar exposure regimen would be increased by depletion of stratospheric ozone. The relevant questions are: (1) How large would the resulting effects be for any given ozone depletion? (2) Would they be tolerable for the depletions anticipated from continued CFM release?

BIOLOGICAL RESPONSES TO SOLAR ULTRAVIOLET RADIATION

The sizable variations in stratospheric ozone concentration that occur naturally should provide a limited base of experience on the biological effects of ozone change. Irregular changes of 20–40 percent, lasting for a few days, accompany weather patterns in middle latitudes. All species that naturally endure sunlight must be adapted to those short variations. Seasonal variations of 20–40 percent also take place in middle latitudes, the high concentrations occurring in the spring, and the low concentrations in the fall. Because these changes are phased with the seasonal change of prevailing solar altitude (in the sky) they merely modify the already-occurring variations in monthly average uv-B radiation at different times of the year.

Longer-term variations, amounting to perhaps 10 percent, swing back and forth in a decade or so. These changes, which must produce roughly 20 percent changes in uv-B when weighted for biological effectiveness, over periods of half a dozen years, more nearly resemble in magnitude the longer-lasting depletions that might be expected from continuing CFM releases at somewhat above present levels. The fact that temperate areas of the world have repeatedly been through such cycles means that a 10 percent ozone decrease does not produce any spectacular effects in plants and animals over a few years time. Detection of such changes would, however, be blurred by the ordinary variations in weather and the effects of factors such as temperature and rainfall. More striking effects of ozone reduction may require longer-lasting or larger depletions than this variation of ±5 percent about the average. Nevertheless, this should not encourage complacency. The reduction of mean atmospheric ozone by CFM release would be superimposed on natural fluctuations. Therefore, periods of low ozone concentration in the course of natural variations, if compounded by a CFM-induced reduction in mean ozone concentration, could result in exposure of ecosystems to levels of uv-B flux not yet experienced.

Coping with solar uv radiation at present-day intensities involves two major strategies for most organisms:

1. Avoidance of excessive solar uv radiation by either physical screening or behavioral means.
2. Tolerance of uv radiation damage through molecular repair mechanisms or through replacement of damaged cells or tissues.

Avoidance of solar radiation by absorption of the radiation in outer body coverings such as feathers, fur, or special pigments in the case of both plants or animals is a common feature of many organisms and can often be inversely correlated with behavior patterns of the animal. For example, Porter (1967 and 1975) showed that many snakes and reptiles that are active during the daylight hours have effective melanin pigment accumulations in outer tissues that prevent uv radiation from penetrating to physiologically sensitive targets. He also found that animals that either lacked these pigments or had them in reduced quantities generally avoided exposure to solar uv radiation through nocturnal behavior. Adaptive increases in pigmentation can also be noted for such effects as the tanning of human skin, where melanin pigments aggregate and form a better barrier to the penetration of solar uv radiation. Nevertheless, the capacity of any particular organism to increase its physical screening against uv radiation is limited. A vivid example is provided in the case of Caucasian man, where, although tanning of the skin aids in reducing sunburn—and presumably other skin lesions, such as cancer—its capability is limited compared with the considerably greater screening provided to the dark-skinned races.

Behavioral avoidance of solar uv radiation is also limited as a protective response to compensate for a decrease in stratospheric ozone. Increased uv radiation flux is not sensed directly by some organisms, including human beings. Such organisms respond instead to the concomitant changes in visible light or radiant heat of normal solar radiation and rely on this response to provide whatever avoidance of excessive uv is necessary. Thus, uv-B *per se* is not the environmental cue that effects the necessary avoidance of uv.

If the ozone shielding were diminished, there would be no accompanying change of the directly perceived cue (light or heat), and consequently no appropriate change or behavior. Even if behavioral avoidance did occur, however, changing the time of day that an organism was active (in the case of animals), or changing its location within an ecosystem, would change its interactions with other species. This might alter such relationships as competition with other species, change in susceptibility to predators, and change in availability of prey or pollinators. Since most species have evolved over millennia to their niche in an ecosystem, it is likely that many such changes in relationships with other species would not be advantageous.

Tolerance of uv radiation damage through the action of molecular repair systems is discussed in Appendix C. Organisms may also tolerate uv radiation damage by replacement of the damaged cells or tissues or in some cases simply by faster reproduction of individuals within a population. For some species there may be a reserve capacity to deal with greater levels of uv-induced damage, but for others not.

Since most species have evolved to cope with present-day intensities of solar uv radiation and the normal fluctuations in uv-B by these various protective mechanisms, one might consider the possibility that evolution could increase the capacity for avoidance or repair, or tolerance (as in the case of man's pigmentation and its latitudinal dependence). However, since the stratospheric perturbation leading to increased solar uv radiation that is envisaged in this report would occur in the context of decades, rather than centuries or millennia, the time available for evolution of increased avoidance or tolerance would make it improbable that many species could achieve heightened uv resistance. Although decades have been sufficient for some notable examples of evolution— such as development of pesticide resistance in insects— to take place, the probability is small that evolution of increased uv resistance would occur in many important species. This is particularly true for species with limited population sizes or slow turnover of generations (Antonovics, 1975).

AGRICULTURAL EFFECTS

Agricultural crop yields in any locale fluctuate from year to year with variations in such things as patterns of temperature, rainfall, and pest predation. The possible consequences of these changes on the prosperity or health of the local population are well appreciated. Fortunately, since the variations are not synchronous over the world, surpluses in one locality can often be used to alleviate shortages in others, or stored surpluses from one year to make up shortages in a subsequent one. The mounting pressure of population on the world's food supply, however, is reducing these margins of safety.

Excessive uv radiation in the vicinity of 300 nm clearly affects plants adversely, as reflected in the reduction of overall growth and photosynthetic activity (Biggs *et al.* 1975). It could, therefore, presumably reduce crop yields in normal agricultural practices. Since an increased solar uv flux in this spectral region would be expected if CFMs continue to be released, the possibility of long-term agricultural decreases must be considered. Even a few percent drop in production of the major food crops, occurring world-wide and lasting for decades, would be an event of economic and social significance.

Information for estimating the magnitudes of crop reductions that might be expected for any given level of ozone reduction is not available. Simple comparisons of plants grown at different latitudes under the different prevailing average ozone shielding—analogous to the studies of latitude effects on skin cancer incidence in humans—will not provide a basis for prediction, since the radiation influence is confounded by simultaneously occurring changes in temperature, moisture, day length, and other factors that affect growth and crop production.

No readily apparent changes in agricultural production have been associated with changes in mean ozone amounting to perhaps 10 percent over periods of a decade or so. However, the rapid introduction of new crop varieties, and the considerable development of agricultural practices in recent decades, combined with the ordinary vicissitudes of weather, would certainly blur the retrospective detection of any moderate effects if they occurred.

Controlled augmentation of the 290–320 nm radiation levels on plants is difficult if one expects to duplicate accurately, on an experimental field-size scale, the change in solar radiation distribution accompanying some specified ozone depletion. On the other hand, approximate simulation of reduced ozone shielding by this means has been produced in the field, as well as in controlled-environment chambers with lamp/filter systems (Sisson and Caldwell, 1975). While these experimental systems suffer from several disadvantages—including the limited irradiation area, which is primarily restrictive for studies of large organism or appreciable populations of organisms—they have nevertheless provided some fairly realistic simulations when properly used and interpreted. In most of the studies carried out on higher plants, solar uv radiation has been supplemented to correspond roughly to a 0.18 atm-cm ozone column of 60° solar altitude. [The normal level usually ranges between 0.28 and 0.36 atm-cm for most of the continental United States between April and September (Hering and Borden, 1967).]

Studies involving both economically important plants and some nonagricultural higher plants have revealed a great heterogeneity of response to the supplemented uv radiation, depending on the experimental conditions and the species involved. For species that are sensitive to the enhanced uv radiation, this uv radiation supplement was sufficient to cause inhibition of plant growth and development, depression of photosynthetic rates, inhibition of pollen germination, and increases in somatic mutation rates (Biggs et al., 1975). However, many species of plants did not appear sensitive to the increased levels of uv irradiation employed. Even among the sensitive species, the degree of damage was quite variable. Although the ability of these levels of uv-B irradiation to cause actinic damage has been clearly demonstrated, quantitative predictions of decreases in crop productivity or alterations of other plant processes in nonagricultural ecosystems are not possible. Furthermore, the response of the more sensitive plant species to lesser degrees of ozone depletion (e.g., a 5 or 10 percent ozone reduction) has yet to be simulated. It is, however, apparent that certain types of damage such as decreased photosynthesis do appear to be accumulative phenomena (Sisson and Caldwell, 1976). Therefore, even with only slightly higher intensities of solar uv radiation, damage in long-lived leaves of sensitive species may accumulate to appreciable degrees in nature. Shifts in the competitive balance between different plant species might also take place under small increases in uv irradiance. A definitive determination of such responses awaits future research.

Possible losses in agricultural animal production due to enhanced uv exposure are also fairly uncertain at this time. Squamous-cell carcinomas of the eye, "cancer eye," in Hereford cattle, which constitute a factor in shortening individual productive life span, are associated with solar uv exposure of unpigmented parts of the lid epithelium (Anderson, 1970; Macdonald, 1975). The effects of increased uv-B intensity on other breeds, and on other species of animals that must remain most of the time in open daylight, are not known. The levels of increased irradiation that could be endured without intolerable skin tumor production or debilitating eye and skin injuries are consequently unpredictable. Qualitatively, the problem is like that of forecasting the effects of increased uv exposure on human populations, but it requires presently unavailable quantitative data for the various individual animal types. Because of the relatively short life span of these animals, a well-planned study, carried out through the next one or two decade-long cycles of ozone variations, and taking full advantage of systematic ozone and uv-B monitoring efforts, should be able to set some limits on the effects of moderate ozone depletions.

EFFECTS ON NATURAL TERRESTRIAL AND AQUATIC ECOSYSTEMS

Apart from the immediate effects of increased solar uv radiation on man and crop production, a broader view must also include the forests, grazing lands, and other nonagricultural regions of the earth as well as the lakes and oceans. All societies are involved to some degree with these large nonagricultural ecosystems. The behavior and quality of watersheds, the supply of pollinating insects for agricultural or wild plants, the balance between agricultural or other pests and their predators, recreational and aesthetic attributes of wild lands as well as the direct production of harvestable materials, such as timber and grazing animals, are all involved in consideration of nonagricultural systems. Man cannot exist in isolation with his cultivated fields. The oceans furthermore represent a large stabilizing resource common to all nations, which in no small part serves as a critical fisheries resource.

These systems are evidently not greatly affected by moderate changes in ozone shielding extending over a few years time, since they are subjected to decade-long cycles in ozone amounts. The effects of larger or more prolonged reductions in the ozone layer superimposed on the natural variations are, however, not certain.

As in the agricultural case, the individual plant and animal species that make up natural ecosystems differ considerably in their sensitivity to uv radiation. However, the properties of the system as a whole could be altered if increases in solar uv radiation affect even fairly few constituent species, perhaps only in subtle ways. Thus, one must be concerned not only with direct changes in populations of a particular organism (which is a matter of concern if it is valuable in its own right) but also with changes in populations of other organisms that depend on, or interact with, the sensitive species through some direct or indirect chain of events. Prediction of the resulting effects of the increase in uv radiation on ecosystems would therefore require mathematical modeling, using a knowledge of the sensitivity and behavior patterns of many of the component species, together with an understanding of how such species interact within the ecosystem. The necessary information and models are mostly lacking at present. In general, only rather small predicted changes in an ecosystem could be viewed with equanimity, since the uncertainty of the prediction would rise with the size of the perturbation, and any large predicted changes, of whatever kind, would run the risk of being regrettable.

Solar uv-B can penetrate into water and when the organic matter content is low will reach depths of the order of 10 m before its effective intensity is reduced to 10 percent of that at the surface. On the other hand, with the higher organic content occurring where biomass production is high, the penetration may be considerably less than 1 m. Phytoplankton (the microscopic green plants that form the base of the aquatic food chain) do appear to be quite sensitive to uv radiation, on the basis of preliminary experiments (Calkins and Nachtwey, 1975). Although many forms of phytoplankton could possibly migrate to greater depths in water to avoid increased intensities, it is questionable whether such an avoidance would provide any great protection in the event of increased solar uv radiation, for the reasons outlined in general terms below. These reasons are:

(1) There is no indication that these organisms sense solar uv-B radiation *per se*, but rather that they probably depend on visible light as a cue for avoiding the normally concomitant uv radiation. (Visible light would not change if the stratospheric ozone were decreased.)

(2) Even if migration to greater depths were to occur, this would result in reduced light to drive photosynthesis and perhaps also in a less amenable environment in terms of other physical factors such as temperature.

In their early life stages, certain fishes and crustaceae tend to frequent shallow waters, where solar uv radiation can penetrate to the bottom. In some cases, egg clusters actually float on the water surface. We may reasonably infer some solar uv radiation damage to DNA, even at present radiation levels, from the existence of very active DNA protorepair systems in cells of marine forms (see Appendix C).

Large and complex natural ecosystems, such as occur in wildlands or in the oceans, though contributing a small proportion of man's direct food needs, are nonetheless extremely important in any consideration of a global perturbation of the ozone layer. Effects on these natural ecosystems are not the less significant for being more difficult to predict, and a strong effort should be made to obtain the necessary information for estimating them in quantitative terms. Within predictability an important element in determining the tolerable levels of ozone reduction is missing.

POSSIBLE EFFECTS OF CLIMATIC CHANGE ON THE BIOSPHERE

A change in the total thickness and distribution of the stratospheric ozone layer as well as infrared activity of the CFMs themselves might have some effect on the energy balance of the lower atmosphere and, thus, an effect on the ground-level climate. However, as was concluded in Chapter 6, neither the net direction nor exact magnitude of this change can be predicted. It appears presently that the principal and more predictable effect of CFM release on climate would be due to the direct intensification of the global greenhouse effect (see Chapter 6). A mean global temperature increase of 0.5°C (0.9°F) has been taken as a working value resulting from continued release of CFMs at the 1973 rate. This discussion is intended to lend some perspective on the significance of such a temperature change for biological effects such as crop yields and changes in natural ecosystems.

The earlier report of the Climatic Impact Committee (1975), which addressed the potential consequences of flight in the stratosphere, considered the biological consequences of changes in the temperature and precipitation for both agricultural crops and nonagricultural ecosystems. A more comprehensive treatment also exists as a monograph resulting from the Department of Transportation Climatic Impact Assessment Program (Bartholic and Jensen, 1975). In these reports, response to climatic change was treated parametrically for a range of changes in temperature (±3°C) and precipitation (±30%). These analyses have served as the basis for this discussion.

Perspective on the effects of climatic change on crop yield is best gleaned from models of crop productivity as a function of weather variables such as temperature and precipitation. Similarly, some indication of the effects of climatic change on the function of nonagricultural ecosystems such as forests and rangelands can be derived from ecosystem-level simulation models that have been assembled in the past decade, largely under the aegis of the International Biological Program. The basis and limitations of these crop and ecosystem models are fully discussed in the earlier reports (e.g., Stewart, 1975). The response of crop yield and other biological processes to factors such as precipitation and temperature is not easily generalized from one crop species to another or from one region to another. These models can thus only serve to provide some indication of the significance of climatic change of a given magnitude.

If the climate were uniformly warmed by 0.5°C and there were no significant changes in the pattern or fluctuation in climatic variables from the present-day climate, the effects of this change on crop yields and natural ecosystems would probably be of small consequence except possibly in areas of marginal crop production or perhaps at the limits of some natural ecosystems. A review of the model calculations for 0.5°C temperature increase for a variety of major crop species, both for North America and representative areas in other continents, suggests changes of usually less than 4 percent of current crop yields. For some crops there is suggested a slight increase in production, and for others, a slight decrease. Even for the same crop such as wheat in the Great Plains states of North America, model predictions suggest slight increases in some states and decreases in other states due to this small warming. The net effect of this global warming on worldwide agricultural productivity is difficult to discern and certainly would be less than the ±4 percent change predicted in particular cases.

The ecosystem-level simulation models for natural systems such as a western coniferous forest and an eastern deciduous forest, both of which are normally well supplied with water, suggest that this warming would result in increased primary production on the order of 5 percent or less.

However, in water-limited ecosystems, such as deserts, this amount of warming might result in small decreases in primary production on the order of 4 percent.

Although the model predictions can be considered to give a rough, first-order approximation of the magnitude of possible biological effects, the rather small suggested changes should not encourage complacency. As was discussed in Chapter 6, models that have yielded the 0.5°C temperature increase as a global average are associated with sizable errors. Furthermore, they do not yield an indication of the possible changes in patterns or degree of fluctuation in parameters such as precipitation, minimum or maximum temperatures, and cloud cover, which could be associated with these changes in global circulation. Shifts in circulation patterns and local changes in climate could be of considerable magnitude. Agricultural or other biological perturbations resulting from these more disruptive local changes might be far from inconsequential.

Furthermore, the biological models employed in the analyses of crop productivity and natural ecosystem response to climate change are also not of sufficient complexity or precision to anticipate many of the interacting effects of changes in climatic variables on plant growth, insect pest population changes, and possible alterations in the balance between plant and animal pathogens and their host species. Many ecosystem processes such as the cycling of inorganic nutrients or the changes in competitive balance between various plant and animal species in response to climatic change, particularly if magnified on a local scale, cannot be anticipated by these mathematical models in their present form. It should also not be forgotten that the effects of even small changes in climate resulting from CFM release must be taken against a background of possible effects of other anthropogenic climatic alterations (such as that due to increased global CO_2 concentrations) as well as the normal vicissitudes of climatic change, which are still so poorly understood. Unchecked CFM release would hardly seem prudent in view of our current inability to assess the ultimate consequences.

REFERENCES

Anderson, D. E. 1970. Cancer eye in cattle, *Mod. Vet. Pract.* 51:43–47.
Antonovics. 1975. Predicting evolutionary response of natural populations to increased uv radiation. In *Impacts of Climatic Change on the Biosphere. Part 1, Ultraviolet Radiation Effects*, pp. 8-3–8-26. D. S. Nachtwey, M. M. Caldwell, and R. H. Biggs, eds. U.S. Dept. Transportation, DOT-TST-75-55, Washington, D.C.
Biggs, R. H., W. B. Sisson, and M. M. Caldwell. 1975. Response of higher terrestrial plants to elevated uv-B irradiance. In *Impacts of Climatic Change on the Biosphere. Part 1, Ultraviolet Radiation Effects*, pp. 4-34–4-48. D. S. Nachtwey, M. M. Caldwell, and R. H. Biggs, ed. U.S. Dept. of Transportation, DOT-TST-75-55, Washington, D.C.
Caldwell, M. M. 1971. Solar uv irradiation and the growth and development of

higher plants. In *Photophysiology.* Vol. 6, pp. 131–177. A. C. Giese, ed. Academic Press, New York.

Calkins, J., and D. S. Nachtwey. 1975. Uv effects on bacteria, algae, protozoa and aquatic invertebrates. In *Impacts of Climatic Change on the Biosphere. Part 1, Ultraviolet Radiation Effects,* pp. 5-3–5-8. D. S. Nachtwey, M. M. Caldwell, and R. H. Biggs, eds. U.S. Dept. Transportation, DOT-TST-75-55, Washington, D.C.

Climatic Impact Committee. 1975. *Environmental Impact of Stratospheric Flight: Biological and Climatic Effects of Aircraft Emissions in the Stratosphere.* National Academy of Sciences, Washington, D.C.

Giese, A. C. 1964. Studies of ultraviolet radiation action upon animal cells. In *Photophysiology,* pp. 203–245. A. C. Giese, ed. Academic Press, New York.

Gordon, M. H. 1975. *Molecular Mechanisms for Repair of DNA.* P. C. Hanawalt and R. B. Setlow, eds. Plenum Press, New York.

Harm, W. 1969. Biological determination of the germicidal activity of sunlight, *Radiat. Res.* 40:63–69.

Hering, W. S., and T. R. Borden. 1967. Ozonesonde observations over North America, Vol. 4, *AFCRL-64-30* (IV), Air Force Cambridge Research Labs, Bedford, Mass.

Kaluskar, A. G., and L. I. Grossweiner. 1974. Photochemical inactivation of trypsin, *Photochem. Photobiol.* 20:329–338.

Lamola, A. A. 1974. Fundamental aspects of the photochemistry of organic compounds; electronic energy transfer in biologic systems; and photosensitization. In *Sunlight and Man,* pp. 38–55. T. B. Fitzpatrick, M. A. Pathak, L. C. Harber, M. Seiji, and A. Kukita, eds. U. of Tokyo Press, Tokyo.

MacDonald, E. J. 1975. Association between cancer eye and solar radiation. In *Impacts of Climatic Change on the Biosphere, Part 1, Ultraviolet Radiation Effects,* pp. 6-3–6-25. D. S. Nachtwey, M. M. Caldwell, and R. H. Biggs, eds. U.S. Dept. Transportation, DOT-TST-75-55, Washington, D.C.

McLaren, A. D., and D. Shugar. 1964. *Photochemistry of Proteins and Nucleic Acids,* Macmillan, New York.

Michalke, H., and H. Bremer. 1969. RNA synthesis in *Escherichia coli* after irradiation with ultraviolet light, *J. Mol. Biol.* 41:1–23.

Munkata, N., and C. S. Rupert. 1974. Dark repair of DNA containing "spore photoproduct" in *Bacillus subtilis, Mol. Gen. Genet.* 130:239–250.

Nachtwey, D. S. 1975. Linking photobiological studies at 254 nm with uv-B. In *Impacts of Climatic Change on the Biosphere. Part 1, Ultraviolet Radiation Effects,* pp. 3-50–3-84. D. S. Nachtwey, M. M. Caldwell, and R. H. Biggs, eds. U.S. Dept. Transportation, DOT-TST-75-55, Washington, D.C.

Porter, W. P. 1967. Solar radiation through the living body walls of vertebrates with emphasis on desert reptiles. *Ecol. Monogr.* 37:273–296.

Porter, W. P. 1975. Ultraviolet transmission properties of vertebrate tissues. In *Impacts of Climatic Change on the Biosphere. Part 1, Ultraviolet Radiation Effects,* pp. 6-3–6-15. D. S. Nachtwey, M. M. Caldwell, and R. H. Biggs, eds. U.S. Dept. Transportation, DOT-TST-75-55, Washington, D.C.

Rupert, C. S. 1975. Enzymatic photoreactivation: overview. In *Molecular Mechanisms for Repair of DNA,* pp. 73–87. P. C. Hanawalt and R. B. Setlow, eds. Plenum, New York.

Rupert, C. S. 1964. Photoreactivation of ultraviolet damage. *Photophysiology* 2:283–327.

Setlow, R. B. 1974. The wavelengths in sunlight effective in producing skin cancer: a theoretical analysis, *Proc. Nat. Acad. Sci.* U.S. 71:3363–3366.

Sisson, W. B., and M. M. Caldwell. 1975. Lamp-filter systems for simulation of solar uv irradiance under reduced atmospheric ozone, *Photobiol. Photochem.* 41:453–456.

Sisson, W. B., and M. M. Caldwell. 1976. Photosynthesis, dark respiration and growth of *Rumex patientia* L. exposed to uv irradiance (288–315 nm) simulating a reduced atmospheric ozone column, *Plant Physiol.*, in press.

Spikes, J. D. 1968. Photodynamic action, *Photophysiology* 3:33–64.

Swenson, P. A., and R. B. Setlow. 1966. Effects of ultraviolet radiation on macro-molecular synthesis in *Escherichia coli. J. Mol. Biol.* 15:201–219

Varghese, A. J. 1972. Photochemistry of nucleic acids and their constituents, *Photophysiology* 7:207–274.

Health Effects

A few decades ago, the western world began to believe that exposure to the sun promotes health and the more exposure, the better. The dangers of general exposure came more clearly in view as sun-seeking habits became more and more widespread. Unfavorable effects range from transient sunburn, through wrinkling and keratoses, to skin cancer. We shall describe these different effects briefly below.

A single 15- to 20-minute exposure to the sun in middle latitudes can produce sunburn. In perhaps more than two thirds of the white population, careful repeated exposures will develop a protective tan. In the remainder, little or no protective tanning occurs, and sensitivity continues, leading to repeated sunburn on exposure.

The longer-term effects of exposure to the sun involve the accumulated dose of exposure, usually over a period of years. Most wrinkling of the skin of the face and hands and all the warty thickenings of the skin called solar keratoses are associated with prolonged or intensive exposure to sunlight. While these effects are far from desirable, their seriousness is limited.

The serious effects of exposure to the sun involve the development in some individuals of skin cancers of three major kinds:

81

—Basal-cell cancers (nonmelanoma)
—Squamous-cell cancers (nonmelanoma)
—Melanomas

Together the first two types of skin cancer add up to the most frequently detected cancer in man. They are also the most easily and most successfully treated human cancers. The quantitative extent to which other agents other than uv exposure cause nonmelanoma skin cancer in white population has not been established. It is, however, believed to be small. Some non-melanoma skin cancer is caused by exposure to arsenic, pitch, and x rays, often in the course of work, sometimes following treatment of skin disorders. This latter group of tumors is found among patients of all degrees of skin pigmentation who happen to be exposed to the irritants. In some less developed countries, most nonmelanomas seem to arise in neglected wounds (Camien et al., 1972; Fleming, 1975).

The report of the Climatic Impact Committee (1975, page 40) stated that: "The available evidence indicates that the spectral sensitivity for skin cancer is similar to either the action spectrum for erythema production or the spectrum for damaging DNA. The two spectra are similar but not quite identical." Nothing has been learned to weaken this conclusion.

Nonmelanoma skin cancer (which rarely leads to death) is presently a serious problem because of disfigurement (frequently minor and infrequently severe) and the significant economic burdens associated with its treatment.

Melanomas, the third mentioned cancer, are a serious life-threatening hazard and are as common as primary malignant brain tumors. Our most recent figures, 1965–1969 (Cutler et al., 1975), show that only two thirds of new melanoma patients survived for 5 years, about the same fraction as for breast cancer.

Again, this fraction is not known quantitatively, but differences in melanoma rates between, for instance, blacks and whites (see Appendix B, page 100) suggest that this fraction is also small in whites. The previous report (Climatic Impact Committee, 1975, page 41) stated that: "Although the evidence associating uv-B with malignant melanomas is not so strong as for nonmelanoma, we believe that the only action spectra that we can prudently use for any quantitative estimate of the potential hazard arising from an increase in uv-B are those given in Figure 4." The evidence collected since that report serves only to strengthen that conclusion, as we shall see below.

In developed countries, a fraction of malignant melanomas not arising by metastasis must come from other causes than ultraviolet radiation.

In terms of human health, then, our concern with the consequences of exposure to the sun:

—Must be heavily concentrated on the production of melanomas (because they threaten life)
—Has to give rather serious attention to the production of other skin cancers (because they are very common)
—Should take other lesser and less well-known effects into account.

Skin cancers associated with solar ultraviolet (uv) radiation, as diseases of the less pigmented peoples, are thus a major concern in the United States, Europe, and Australia—and to emigrants from these regions to other parts of the world. (The, as yet uncertain, effects on plants, animals, and climate will be the main consequences of CFM release for the more pigmented peoples.)

GENERAL MECHANISM OF SOLAR SKIN-CANCER CAUSATION

It is generally accepted that changes in individual molecules of the body's DNA (deoxyribonucleic acid) are most often the key step in cancer production. Accordingly, we expect that sunlight-induced cancers probably arise by the action of sunlight on DNA or compounds of analogous sensitivity to different wavelengths. Ultraviolet radiation of wavelengths below 290 nm does not reach the ground, and uv radiation of wavelengths above 320 nm does not attack DNA seriously. Accordingly, whenever sunlight stimulates skin cancer, DUV (defined as uv-B in the wavelength range 290–320 nm weighted in accordance with its effectiveness in altering DNA) should be the portion of the uv that is effective. (This has been confirmed in animals for nonmelanoma skin cancer.)

OTHER EFFECTS

Against unfavorable effects of uv-B ranging from sunburn to skin cancer, we must place in the balance one favorable effect of widespread importance:

—Conversion, in the skin, of 7-dehydrocholesterol to vitamin D_3 thus preventing rickets.

This action is important to all of us in the world but fortunately does not require large doses of uv-B (winter exposures suffice, even in quite high latitudes). Small percentage changes in DUV are thus not important.

Attention is often directed to sunlight-caused cataracts, but these are caused by uv-A, not by uv-B, and will be unaffected by ozone depletion.

FORMS OF EVIDENCE

Beyond these general considerations, evidence for sunlight as a major or contributing cause of skin cancers takes five forms:

—Differences in incidence or mortality with latitude, corresponding generally to differences in exposure;
—Locations of skin cancer on the body with regard to different degrees of exposure;
—Differences in the incidence or mortality between occupations corresponding to differences in amount or character of exposure;
—Changes in incidence or mortality over time, as plausibly related to changes in exposure behavior;
—Experimental confirmation in animals (so far only for nonmelanomas).

What patterns ought we to see in the first four of these forms if

1. Solar radiation alone causes skin cancer—operating through total accumulated dose—modulated mainly by individual differences in sensitivity and in total exposure?
2. The effect of solar radiation is not solely determined by total accumulated exposure, so that an individual's exposure experience may be important in other ways than total exposure? (Solar radiation may be one of several preconditions for skin cancer, and individual sensitivity will still be involved.)

Where total accumulated dose matters:

—We ought to see a general increase of incidence as the total DUV received increases. [The geographic pattern of incidence can be expected to be modulated by differences in ethnic composition (northern Europeans, for instance, appear to be more sensitive than southern Europeans] and by geographic differences in behavior patterns, so that an exact determination of incidence by the total amount of DUV reaching the ground is not to be expected.]
—The skin cancers should be heavily concentrated on the parts of the body most exposed to the sun.
—Occupations such as farming and street repair, which involve heavier accumulated exposures, should show higher incidences.

—Changes in incidence or mortality in a given locality with time will be slow, since the total exposure of those most exposed is only slowly changing, unless they start to take deliberate protective measures.

These are, of course, the characteristics shown (see Climatic Impact Committee, 1975, pages 36–40) by nonmelanomas.

In the second case, where dependence on exposure is more complex:

—We should expect an increase of incidence and mortality similar to that in the first case, as latitude decreases, but somewhat more perturbed by other factors.

—The pattern of distribution of skin cancers over the body would have to have some sensible relation to exposure, not necessarily being greatest in the most exposed regions but certainly avoiding the least exposed regions.

—The dependence on occupation could be quite complex (since patterns of intermittences of exposure could matter) so that certain occupations with lower total doses might show higher incidences or mortalities. (Occupation-related exposure to other preconditions might have a similar effect.)

—Substantial changes in incidence or mortality over time could occur, especially if there are large changes in exposure pattern, for example, due to changes in dress or in extent or character of recreational exposure.

We shall see, in the remainder of this chapter, that what limited information we have about melanoma incidence and mortality is quite consistent with the second case, where dependence on exposure is more complex and other contributing factors may be involved.

STRENGTH OF EVIDENCE

Granting that, what can we say about the overall strength of the evidence implicating solar uv radiation in the causation of skin cancer and thus leading us to expect increased skin cancer when the amount of DUV reaching the ground increases? For nonmelanomas three kinds of evidence—latitude dependence, body location, and occupational differences—all combine to point closely to exposure to the sun as a prime cause and to increased incidence as a quite certain consequence of increased DUV.

The situation for melanomas is somewhat different. The latitude dependence appears well established. While melanomas are not concentrated on the most exposed regions of the body, they do appear to avoid the least exposed regions, and differences in patterns of locations between sexes correspond to the differences in exposure to sunlight (for instance, more melanomas on the legs of females). Occupational differences are fairly large, and, while they do not correspond to differences in total exposure, they do seem to correspond to differences in patterns of exposure. Changes in mortality and incidence with time are large and can reasonably be associated with changes in exposure behavior.

This leaves us with a qualitatively well-established relation of solar uv radiation to nonmelanomas and a well-founded anticipation of increased nonmelanoma incidence if the DUV reaching the ground were to increase.

The contribution of solar uv radiation to the production of nonmelanoma skin cancers is thus a well-established health hazard of some magnitude and ought to be responded to accordingly.

Less firmly, but we believe persuasively, there is a large likelihood that solar uv radiation contributes to the induction and/or development of melanomas and that an increase in the DUV reaching the ground will induce an increase in melanoma incidence and mortality. (Such an increase may be small compared with the increases now going on, presumably as a result of changes in patterns of exposure.)

If, as we believe, DUV plays an important role in many melanomas, the dose–response relation (which may really be the dose-intermittency-response relation) is complicated and involves many other variables.

To summarize, we believe that the relation of solar uv ultraviolet to melanoma ought to be taken as a likely health hazard of significant size and responded to accordingly.

LATITUDE DEPENDENCE

Let us turn next to the details of the evidence.

The dependence of both incidence (occurrence) and mortality on latitude has been clearly documented for melanoma skin cancer as has dependence of incidence on latitude for nonmelanoma skin cancer (Climatic Impact Committee, 1975, pages 36–41, from which Figure 2 is reproduced). Higher exposure to DUV is reflected in more skin cancer.

As we move toward the equator, the sun is more nearly overhead, and the total amount of sunlight increases at all wavelengths. This increase is by a larger factor as we move from the visible through the uv-A and into the uv-B region. In the uv-B, the controlling factor is absorption by ozone, which increases sharply toward the poles, both as we move under higher ozone amounts and as the sun departs from the zenith. Thus, the accumulated dose of uv-B—or, more particularly, of DUV—is likely to be greater for those living in lower latitudes.

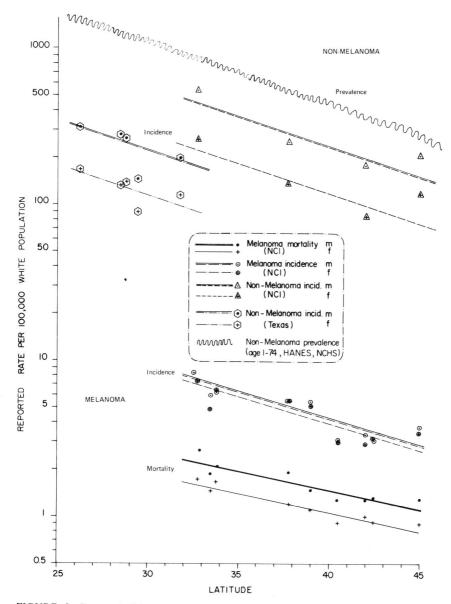

FIGURE 2 Reported skin-cancer rates among whites as a function of latitude. [Sources: melanoma mortality from Mason and McKay (1974), melanoma incidence from National Cancer Institute (1974), nonmelanoma skin-cancer incidence (NCI) from Scotto *et al.* (1974), nonmelanoma skin-cancer incidence (Texas) from Macdonald (1974), and prevalence of nonmelanoma skin cancer based on preliminary data from the Health and Nutrition Examination Survey of the National Center for Health Statistics (McDowell, 1974).]

The effect of more sun may well be partly compensated for by changes in living habits that lead to a smaller fraction of time in the strongest open sun or somewhat enhanced by differences in the opposite direction. Measures of the extent of this compensating (or enhancing) effect do not seem to exist.

BODY LOCATION OF NONMELANOMAS

Most nonmelanoma skin cancers occur

—On light-skinned persons, particularly those who repeatedly burn with little or no tanning
—On exposed areas, especially on head, neck, arms, and hands

[See p. 185 of Climatic Impact Committee (1975) and references cited there for pictures of the distribution.]

While, as noted above, a small fraction of nonmelanoma skin cancers are clearly due to other causes, what we know about nonmelanoma skin cancer suggests that the accumulated DUV dose (built up over years) is the crucial factor.

BODY LOCATION OF MELANOMAS

Until recently, the relation of sunlight exposure to the development of melanoma has been hard to interpret because the pattern of melanomas over the different parts of the body is different from the pattern of nonmelanoma skin cancer.

Population-based incidence data (incidence rate per unit area of skin) from Queensland show significant excess melanomas on the generally (or occasionally) exposed sites of face, leg, neck, and arm in women and face, ear, neck, and back in men. They also show proportionally fewer tumors in parts of the body that are virtually never exposed (Elwood and Lee, 1974).

According to our population-based incidence figures (Third National Cancer Survey, 1975), U.S. white men also have most melanomas on the face and head and trunk. Women have rates similar to men on the face and head and trunk but tend to have higher rates on arm and leg. Limitation of available data, to for example, face and head combined, limits the extent to which these observations can be interpreted.

There are clear variations in incidence as latitude changes within the United States. Melanomas on totally exposed sites show the most, and those on rarely exposed sites the least, latitude dependence (Scotto, 1976). Melanomas of the partially exposed (relatively exposed) sites are increasing in incidence over time at a faster rate than those on sites that have always been totally exposed (Lee, 1976).

Most melanomas seen in a cooperative study involving four large and widely separated U.S. clinical centers (534 cases to date) occur on lightly covered or occasionally uncovered regions of the body. A similar pattern of location of primary melanoma has been reported by Braun Falco (1974). Very few have been seen (either in males or females) on the regions routinely covered by bathing suits. The difference between sites of occurrence on males and females are believed by these investigators to reflect different covering practices. Thus, women who usually wear dresses have many more melanomas on the potentially exposed part of the legs than do men who wear usually trousers. Men, who expose their trunks to the sun during recreation, have more melanomas on their chests than do women. It is highly probably that the few melanomas seen in areas virtually always covered by relatively heavy clothing are due to other causes than uv light.

There is a clinical impression that a majority of cases of melanoma are seen in younger middle- and upperclass males and females who pursue active outdoor recreational activities but spend their working hours indoors.

In current urbanized U.S. society, a general picture of most melanomas occurring on lightly clothed or "on-and-off" clothed body areas on persons not in indoor occupations is quite compatible with some mechanism associated with the DNA sensitivity and with the observed latitude dependence of melanomas on these particular sites.

OCCUPATIONAL DIFFERENCES

The only large-scale analysis of mortality from malignant melanoma was made by the registrar general in England and Wales for the years 1949 to 1953 and 1959 to 1963 (Lee, 1976). As might have been expected (because of higher total exposures to the sun), the rate of nonmelanoma skin cancer was higher among unskilled and skilled than among professional and managerial-type workers. It was quite the opposite with malignant melanomas, where the higher mortality was concentrated in younger professional and managerial workers, confirming the clinical impression mentioned above and quite consistent with a high efficacy of recreational exposure. The most recent British incidence figures confirm higher rates of melanoma in the middle class *(Registrar General's Statistical Review,* 1975). While we need much more information to understand why melanomas are in some way associated with a middle-class way of life and with modest and moderate exposure to sunlight, the weight of evidence is consistent with their being in some way a consequence of exposure to uv light (see below).

CHANGES WITH TIME, PLAUSIBLY RELATED TO CHANGES IN EXPOSURE BEHAVIOR

Some estimate of the tremendous impact of changes in lifestyle (leading to increased exposure) can be derived from careful studies of the experiences of young white women before and after the Second World War, a time when widespread changes in dress and behavior began to occur. From 1911 to 1940, approximately four out of each million United Kingdom women aged 15–44 died each year from skin cancer. As, at these ages, (a) deaths from squamous-cell cancers are virtually unknown and (b) skin carcinomas are confined to males, these deaths can be confidently assigned to melanoma. In other words, in a country (latitude above 50) with limited sunlight, there was a background rate of almost four melanomas per million among young women who did not deliberately expose themselves to sunlight.

In Connecticut, women of the same ages had a melanoma incidence rate of 3.6 per million between 1935 and 1939. Given low survival, this incidence rate compared well with the pre-1940 U.K. mortality. In both countries, there was a rapid post-1940 increase to a mortality rate of 7.8 in the United Kingdom in 1966–1970 and an incidence rate of 45.0 in Connecticut in 1970–1972. The difference between twofold and tenfold increases, in the United Kingdom and Connecticut (see Table 6), is consistent with the different increases of sun exposure associated with similar changes in lifestyle. (Other explanations could, of course, be offered.)

TABLE 6 Skin Cancer in Females 15–44 Years of Age

United Kingdom[a]		Connecticut Melanoma[b]	
Year	Mortality Rate (per Million)	Year	Mortality Rate (per Million)
1931–1935	3.7		
1936–1940	3.7	1935–1939	3.6
1966–1970	7.8	1970–1972	45.0

[a]Essentially all melanomas (see text).
[b]Data from Connecticut Tumor Registry.

ARE MELANOMA INCIDENCE AND MORTALITY REASONABLY CONSISTENT WITH A MAJOR ROLE FOR SOLAR ULTRAVIOLET RADIATION?

Earlier we set down the general outlines of the pattern of evidence to be expected if total uv dose was not the sole critical factor in solar-related cancers (and uv-B might be a contributing but not sole cause) and pointed out that melanoma behavior fitted that pattern. We have now reviewed the details of the evidence concerning melanoma and ought to inquire somewhat further as to the reasonableness of a major role for solar uv radiation in melanoma.

We do not intend to propose specific mechanisms—to do this would be inappropriate at this time. We do try to point out some classes of mechanisms that would allow varying degrees of covering (sometimes total, sometimes none or partial, the latter from semitransparent clothing) of some part of the body or intermittent recreational exposures to increase the incidence of melanoma.

We know that persons of intermediate sensitivity to sunburn develop it after initial exposures but eventually tan. On a generally exposed portion of the body, or when occupational exposure occurs daily throughout the year, the annual re-enhancement of protective tan would often take place in the spring when total daily doses are low. Thus no sunburn would be produced by the heavy doses of summer—and it is possible that any melanoma-inducing effect of these doses would be greatly reduced. On a partially exposed portion of the body, or when recreational exposures are both intermittent and seasonably concentrated, the annual re-enhancement of tan would be likely to wait until summer and result from much more intensive doses and involve considerable sunburn. It is possible that any melanoma-inducing effect would then be stronger although the total accumulated dose is less. Similar behavior could arise when clothing is sometimes completely uv-absorbing and sometimes not.

Other possibilities for underneath clothing—experience with sunburn patients shows that significant amounts of uv radiation do penetrate women's clothing—would include combinations of uv exposure with other unknown factors, such as the state of hygiene or even mechanical friction. If photorepair occurs in man (it seems not to be yet documented in mammals), it, too, might provide an explanation for more response on incompletely covered regions.

What about the effects of changes in habits? What if people, particularly those apparently most likely to be affected (sunburn easily, never tan) were to expose themselves less to the sun? How would we expect such changes to be reflected in melanoma rates and incidences?

Suppose, for example, that total exposure (or total exposure in a given pattern) is the critical factor and that incidence is proportional to total dose. Then, if all susceptibles reduce their accumulated exposure time (allowing for hour-to-hour and season-to-season changes in DUV) by 20 percent, thus receiving 80 percent of what might have been the dose, and if, at the same time, the ground-level DUV increases by perhaps 25 percent, the two changes could compensate, and the incidence of—and mortality from—melanomas would be the same. This might at first seem to say that the habit change "took care" of the increased DUV. Much more realtistic, however, is the comment that if we could have kept the habit change but not increased the DUV we would have had perhaps 20 percent fewer melanomas.

Desirable habit changes, especially by susceptibles, can be very worthwhile, affecting general levels of incidence and mortality. But the losses from more DUV at ground level are likely to involve increases in skin cancer, by a fixed fraction, whatever base level we may be able to reach by habit change.

FORECASTING THE EFFECTS OF DUV INCREASE

If uv light, which varies with latitude, is responsible for most melanomas diagnosed in the United States, the observed relationship between DUV and the incidence or mortality of melanoma in the United States can be reasonably used to forecast the effects of change in DUV intensity on both non-melanoma incidence and melanoma mortality.

Statistical studies by the National Research Council's Panel to Review Statistics on Skin Cancer were reported in the report of the Climatic Impact Committee (1975, pages 177–221). More recent studies include those by Scotto *et al.* (1976) and Green *et al.* (1976). These studies have a number of characteristics in common, namely:

—The dependence of both number of cases (incidences) of either non-melanoma or melanoma and number of melanoma deaths (mortality) on geographic location shows, generally, more cases—and more deaths—where the accumulated DUV at ground level is higher, which is roughly the same as being nearer to the equator.

—This effect cannot account for all the observed differences in incidence and mortality, something that is only to be expected in view of differences in susceptibility and habits of exposure.

—These analyses do not—and could not be expected to—provide independent and conclusive evidence that increased DUV, rather than other variables that increase smoothly toward the equator, is the causative factor for either melanomas or nonmelanomas.

Scotto (1976) and Green *et al.* (1976) differ in their chosen dose–response relations, the one using logarithmic rates proportional to dose and the other rates proportional to dose to a power. Neither study makes allowance for varying behavior of cohorts, which may or may not be important in this aspect of their contexts. Both studies, if taken at face value, would give percentage increases in nonmelanomas up to twice the percentage increase in DUV.

The proper reliance on these statistical studies is a basis for estimating the increased effect that results from more DUV and not an aid to establishing cause and effect. As noted earlier, it is our general understanding of agent-caused cancers and the direction—not the amount, not the detailed behavior—of changes in incidence and mortality as we move from poles to equator that leaves us seriously concerned that DUV exposure is related to skin-cancer induction.

To say this does not deny the probable importance of other factors, acting either alone or in conjunction with DUV. (The localization of melanoma in lightly/irregularly covered portions of the body may illustrate the importance of at least one factor that combines with DUV. Geographic differences in melanoma mortality among places with similar accumulated DUV doses show the importance of other factors, some of which may act independently of DUV, although others seem likely to act concurrently with it.)

Given our general information about human behavior as it depends on latitude, some will anticipate that the average residents of lower latitudes accept a smaller fraction of the available DUV dose than do those average residents in higher latitudes. (Because of the much larger accumulated dose available in lower latitudes, lower-latitude residents would receive much greater total doses.) The difference in melanoma mortality—or either in nonmelanoma or melanoma incidence—between the two latitudes would then be less than that corresponding to the difference in the DUV dose, since behavior differences will partially compensate for the enhanced dose in lower latitudes. Others would anticipate residents in lower latitudes accepting a higher fraction of the available DUV dose, thus enhancing, rather than partially compensating for, the altitude difference in DUV dose. (There will also, probably, be some compensation, on the average, due to differences in the proportion of residents of highest sensitivity, mainly reflecting ethnic composition.)

What if the DUV reaching the ground increases, so that current intensities at higher latitude now equal previous intensities at the lower latitude? The accumulated intensities reaching the ground at the higher latitude will build up, somewhat slowly, toward those previously occurring at the lower latitude. The behavior patterns may also change slowly, but probably not to match previous patterns at the lower latitude, since if only DUV is increased, the only cue for a greater avoidance to the sun will be slight increases in sunburn, while temperature, brightness, etc. will be unchanged. The change in behavior will probably be slower than the change in accumulated intensity. (And any changes in ethnic composition will be, relatively, still smaller and still slower.)

As a result, the accumulated doses for individuals at the higher latitudes will rise toward values that may be slightly above those originally found at the lower latitude. This will occur if behavioral compensation (and ethnic composition compensation) will be even less complete than those now in effect, if these are partially compensating. Thus, if we use today's differences, found in statistical studies, between latitudes, both in DUV and skin-cancer rates, to forecast the eventual changes (after 1, 2, or 3 decades) in skin-cancer rates flowing from an increase in DUV at one latitude, we will somewhat, but probably not greatly, underestimate the effects of the increased DUV. We may, also, of course, if present behavior enhances latitude differences, rather than partially compensating for them, somewhat overestimate these effects.

We judge the assessment of the expected increases in deaths and numbers of cases due to a given increase in DUV to be fairly estimated (though perhaps slightly underestimated or, conceivably, slightly overestimated) by the results of statistical studies relating mortality or incidence to the different accumulated DUV doses in different geographical locations.

CONCLUSION

We find that an increase in melanoma deaths is likely, but not certain, to occur as a consequence of a continuing increase in the rate at which DUV received at the ground accumulates. Such a melanoma increase, if it occurred, would be delayed, beyond any delay in the DUV increase, while the accumulated dose builds up in the individuals. A 7 percent ultimate reduction in ozone, with a consequent 14 percent ultimate increase in DUV accumulation rate, might be expected, if most melanoma deaths are solar uv radiation related, to produce a somewhat smaller percentage increase (less than 15 percent) in melanoma deaths. Thus, a few hundred deaths per year would be expected after all delays have taken place. (With a constant rate of CFM release, the increase in DUV arriving at the ground would first reach two thirds of its ultimate level after 80 years, so that more than a century of continued release at a constant level would be required for three fourths of such an effect to be manifest.)

We have done nothing to modify or improve the estimates for increases in nonmelanoma incidence given in the report of the Climatic Impact Committee (1975, pages 41-45) although we look forward to continuing improvements in such estimates, particularly through well-planned data collection, preferably including such covariates as skin type.

REFERENCES

Braun Falco, O. 1974. Maligne Melanome der Haut aus dermatologischer sicht, *Chirurg* 45:345–356.

Camian R., A. J. Tuyns, H. Sarrat, C. Quenum, and I. Faye. 1972. Cutaneous cancer in Dakar, *J. Natl. Cancer Inst.* 48(1):33–49.

Climatic Impact Committee. 1975. *Environmental Impact of Stratospheric Flight: Biological and Climatic Effects of Aircraft Emissions in the Stratosphere,* National Academy of Sciences, Washington, D.C.

Cutler, J., M. H. Myers, and S. B. Green. 1975. Trends in survival rates of patients with cancer, *New Engl. J. Med.* 293:122–124.

Elwood, J. M., and J. A. H. Lee. 1975. Recent data on epidemiology of malignant melanoma, *Seminars Oncol.* 2(2):149–154.

Fleming, I. D., J. R. Barnawell, P. E. Burlison, and J. S. Rankin. 1975. Skin cancer in black patients, *Cancer* 35(3):600–605.

Green, A. E. S., G. B. Findley, Jr., K. F. Dlenk, W. M. Wilson, and T. Mo. 1976. The ultraviolet dose dependence of non-melanoma skin cancer incidence. Unpublished data.

Lee, J. A. H. 1976. University of Washington. Personally communicated data. Occupational Mortality Tables. 1971. H.M.S.O., London.

Macdonald, E. J. 1974. A statement on skin cancer incidence and sun exposure, manuscript.

Mason, T. J., and F. W. McKay. 1974. *U.S. Cancer Mortality by County: 1950–1969.* DHEW Pub. No. (NIH) 74–615. U.S. Govt. Printing Off., Washington, D.C.

McDowell, A. 1974. Nat. Center for Health Statistics. Personal communication.

National Cancer Institute. 1974. The Third National Cancer Survey Advanced Three Year Report: 1969197l Incidence. Preprint.

Registrar General's Statistical Review of England and Wales for the Three Years 1968–1970: Supplement on Cancer. 1975. Table E, p. 29.

Scotto, J. 1976. Melanomas of the eye and other noncutaneous sites: epidemiologic aspects, *J. Natl. Cancer Inst.* 56(3):489–491.

Scotto, J., A. W. Kopf, and F. Urbach. 1974. Nonmelanoma skin cancer among Caucasians in four areas of the United States, *Cancer* 34:1333–1338.

Scotto, J., T. R. Fears, and G. B. Gori. 1976. Measurements of ultraviolet radiation in the United States and comparisons with skin cancer data, DHEW Publ. No. (NIH) 76–1029.

Third National Cancer Survey 1969–71: Incidence Data. 1975. National Cancer Institute Monograph 41.

Appendix A
Control Actions

The discussion that follows is important because it illustrates at least two routes toward the reduction—but NOT elimination—of the use and release of CFMs. The specific routes considered here may—or may not—prove to be the most feasible, but they show that control is feasible and illustrate some of the difficulties that will have to be overcome.

The control of chlorofluoromethane releases to protect the ozone layer and thus human health needs to be focused, for the next few years, on two compounds now in large-scale use, F-11 (CFCl$_3$) and F-12 (CF$_2$Cl$_2$). Replacement of these compounds in some uses, not only by nonozone-attacking materials but also by less-ozonethreatening CFMs, such as F-22, may become desirable. As the use of such other less-ozone-threatening CFMs increases, the time may come when analogous restraints on their use will become appropriate. For the present, however, the term "CFM" in this report is intended to be restricted to F-11 and F-12.

In view of the present importance of certain uses of CFMs to human activity, including human health, it is not appropriate, however, to eliminate all release of CFMs to the atmosphere at any date that can now be envisaged.

It is important to stress that both the ozone reduction and consequent effects caused by release of CFMs will be spread over many decades. Accordingly, the benefits of the reduction in CFMs released in any year will be spread over correspondingly long time.

REGULATION

In view of the general considerations mentioned above, any regulatory action should be selective, and almost certainly occur in stages, so that one

97

use after another is restricted, at times expending over many years, and so that some uses are left untouched.

A few statements need emphasis:

1. The varying importance to mankind of diverse uses of CFMs and their varying contribution to ozone depletion, and hence to human skin cancer, make it entirely inappropriate to take the same actions for all uses.

2. These varying importances and varying impacts combine with varying engineering difficulties in the replacement of CFMs for specific uses to make it often inappropriate, even when the same action should eventually be enforced for two uses, for this to become effective on the same date.

3. Emphasis on balance of health effects and on engineering feasibility does not exclude attention to economic disruption.

4. Future increase in scientific knowledge, which will be accelerated to the extent that our earlier recommendations are implemented, will make detailed decisions both easier and more precise as time goes on.

Accordingly, we suggest more specifically that, starting either immediately or when the decision to regulate has been taken on appropriate evidence:

(a) Mechanisms be developed for the selective control of the diverse uses of CFMs, perhaps by general licenses applicable to each permitted use, which would initially exist for all present uses and would be selectively restricted or withdrawn. [For simplicity only, the suggestions that follow will refer to "licenses" and "cancellation of licenses" without prejudice to the mechanism(s) to be finally chosen.]

If after a year or two of gathering increased knowledge, it is decided that we need to proceed with selective regulation of CFM use, we suggest that:

(b) A firm policy be adopted that licenses will be withdrawn, after a suitable delay, from uses for which reasonable nonpolluting substitutes become available.

(c) In judging the reasonableness of a substitute, attention be given to the human importance of the uses, with more important uses requiring more closely equivalent alternatives before cancellation of the license for that use is contemplated.

(d) In determining an effective date for the cancellation of the license for a use, serious and careful consideration be given to the difficulties, including those of engineering and economic readjustment, involved in the change.

(e) Licenses to use, wherever appropriate, be made conditional on changes in structural design or other factors that will reduce the release of CFMs to the atmosphere.

(f) Where appropriate, restrictions on service and construction practices be imposed by other mechanisms (other than broad licenses by use or their equivalent).

We would find it appropriate if, among the many measures that would have to be considered, the five isolated actions listed below, for example, were taken with greatly different effective dates, starting after perhaps a year after the decision and delaying the later items in the list successively longer.

(1) Effective cancellation of the license to use (manufacture) CFMs for most aerosols (although continuation of some special uses, at least for an additional time and possibly indefinitely, would probably be appropriate). (The British report *Chlorofluorocarbons and Their Effect on Stratospheric Ozone* says, page 14, "Some CFC 11 and CFC 12 containing aerosols, such as many of those used for medical and pharmaceutical purposes, and for which there is no ready satisfactory alternative, can be classed as essential. However, many of the uses of aerosol products, particularly for cosmetics and toiletries, are items of convenience and not of necessity;. . . .")

(2) Requirement that all repairers of automotive air conditioners use equipment and procedures to retain used CFMs and return them for reprocessing.

(3) Requirement that new automotive air conditioners be so constructed as largely to avoid the present releases. (This action would have to be based on a careful study of the relative importance of losses at the seals, diffusion through flexible elastomer hoses, and losses at joints.)

(4) Effective cancellation of the license to use (manufacture) automotive air conditioners using CFMs. (This might never be appropriate, if other actions sufficiently reduce the total CFM releases.)

(5) Actions to curb release associated with household air conditioners, industrial coolers, and industrial refrigeration would appropriately wait their turn, again with action on practices preceding any cancellation of use, if that ever becomes appropriate.

We note that, if present estimates of relative releases are at all correct,

(6) Effective cancellation of the license to use (manufacture) home refrigerators using CFMs would never become appropriate.

We must emphasize the importance to all of us, consumers and manufacturers alike, of adequate time intervals between decisions to cancel licenses and the effective dates of such cancellations. No one's interest, we believe, is served by cancellations without appropriate delays. There should be adequate notice, within a procedure that recognizes the differing human values of different uses and that is intended, over a reasonable span of years, to diminish substantially the release of CFMs both in the United States and worldwide.

HEAVY TAXATION AS AN ALTERNATIVE

Sufficiently heavy taxation would undoubtedly decrease the use and release of CFMs. Whether control by taxation, where the relative impact in different countries is hard to compare, is adaptable to worldwide control of a worldwide phenomenon seems to us quite doubtful.

If control by tax is to have the desired effect, taking account of the value of various uses as well as of the amount used/released, there would have to be a system of rebates for more desirable uses. Thus, initial fillings of household refrigerators, for instance, should, receive a rebate of at least 90 percent. (No rebate for replacement fillings or fillings in connection with repairs would be both desirable in itself and a simplification in administration.)

A large rebate on used CFMs returned for reprocessing could, in effect, be provided by eliminating any tax on reprocessed materials. Local or small-scale reprocessing would thereby not be inhibited.

The extent of tax required to make a substantial decrease in CFM use/release is hard for us to gauge but would certainly have to be substantial. Perhaps three to five times the present wholesale cost would be a starting point. Taxes of this magnitude inevitably produce subtle and ingenious attempts at evasion, as well as crude ones. Difficulties in enforcement need to be considered in any decision to depend on taxes for control.

Any tax intended for control should incorporate provisions automatically raising the tax if the desired reduction in use/release is not obtained.

Appendix B
Further Detail on
Malignant Melanoma

U.S. RATES OF INCIDENCE

The best available estimates of current incidence rates of melanoma come from the Third National Cancer Survey (1975), based on counts during the years 1969-1971. Table B.1 gives both crude and age-adjusted rates for the entire population and for sex-color groups.

U.S. FIVE-YEAR SURVIVAL

According to the National Cancer Institute's *End Results in Cancer* (Report No. 4), 1972 and *Recent Trends in Survival of Cancer Patients 1960-1971* (1974), the 5-year adjusted survival rate is 67 percent for malignant melanoma patients diagnosed between 1965 and 1969.

TABLE B.1 Average Annual Incidence Rates per 100,000 U.S. 1969-1971

Population	Rate Crude	Age Adjusted (1970)
Total	4.1	4.2
White Males	4.4	4.6
Females	4.6	4.4
Black Males	0.7	0.9
Females	0.6	0.7

101

PERCENTAGE OF ALL CANCERS

Melanomas make up 1.4 percent of all cases of cancer diagnosed between 1969-1971 and about 1 percent of all cancer deaths in 1970. These figures come from the Third National Cancer Survey (1975) and the U.S. Vital Statistics Reports for 1970.

Applied to a current estimated population of 200 million, the results of this survey would indicate 8400 new cases each year. The survival rates quoted above suggest that 2772 of these new patients will die within 5 years.

OTHER SITES OF COMPARABLE INCIDENCE

Melanomas can be viewed from the perspective of other cancers found with similar frequency. The rates of relative survival and of number of estimates deaths from the cancer within 5 years are included with the incidence rates of these sites in Table B.2.

COHORT EXPERIENCE

Lee and his colleagues (Lee, 1975, 1976; Elwood and Lee, 1975) have shown definite changes in the malignant mortality experience of United Kingdom birth cohorts since the late 1800's. Careful scrutiny of the limited collection of years represented in the Connecticut tumor registry for white women and a recent published review of experience in Columbia-Presbyterian Medical Center (Cosman et al., 1976) suggest that an increasing lifetime incidence also exists in the United States.

TABLE B.2 Cancer Sites with Comparable Incidence Rates

Site	Annual Age-Adjusted[a] Incidence per 100,000 (1969-1971)	Year Relative Survival (1965-1969) (%)	Estimated Current Deaths from Cancer within 5 Years (Rounded)
Melanoma (skin)	4.2	67	2800
L.R.C.S.[b]	4.8	32	6500
Brain	4.5	29	7200
Larynx	4.2	61	3300
Thyroid	3.6	84	1200
Multiple myeloma	3.6	16	6000
Esophagus	3.4	4	6500

[a]1970.
[b]Of lymphosarcoma and reticulum-cell sarcoma.

TABLE B.3 Primary Melanoma of Skin (in White Persons)

Type of Melanoma	Median Age (yr)	Specific Sites[a]	Rate of Development	Appearance
Lentigo-maligna melanoma	70	Face, neck and hands	Slow: 5-20 yr.	Predominantly flat spot 2-20 cm in size, with *irregular* borders and with raised portions throughout.
Superficial spreading melanoma	56	Face, neck, upper trunk, and lower legs (in females)	Moderately slow: 1-7 yr	The brown or black color is notched; or speckled; or speckled; blue, white, and red; or both
Nodular	49		Rapid: months	Isolated small (3.0 cm) nodule with *smooth* borders; color uniform blue-black

[a]All three types occur either on the exposed parts of the face, neck, and hands or on the relatively exposed areas of the chest, back, and legs. Only a few lesions are seen on covered areas such as the breasts of females, bathing trunk areas of males, and bathing suit areas of females.

GEOGRAPHIC VARIATION

Within the United States, incidence doubled for each 10° decrease in latitude (Climatic Impact Committee, 1975).

TYPES OF MELANOMA

Malignant melanoma arising in the skin is not one but three different specific types of cancer of pigment cells (see Table B.3). The differences are related to the age of onset, the clinical appearance, the cellular pattern (histology), the rate and manner of growth in the skin, and the survival rate.

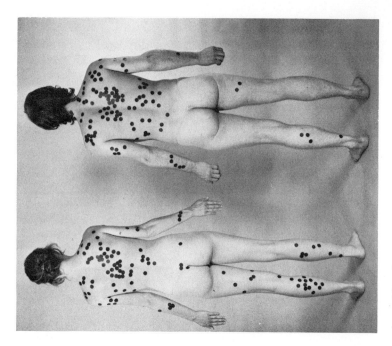

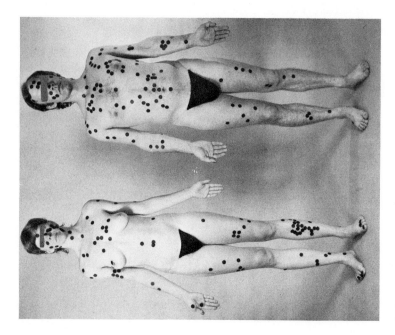

FIGURE B.1   Exhibit of localization of melanoma skin cancer in 534 males and females.

104

The relative occurrence of the three types, as seen in different kinds of hospitals, appears to be quite different, with superficial spreading melanoma as high as 80 percent in general hospitals and reported to be as low as 10 percent in specialized cancer hospitals. Because two of the three types of melanoma may be present as an identifiable cancer for several years before invading deeply in the skin, it is possible to remove these early lesions surgically before deep invasion occurs and by this treatment have a chance of curing the disease. This, of course, depends on early detection of primary melanoma by patient or physician, which might reasonably be achieved by an intensive public educational program, perhaps even by as simple a program as adequate publicity for the following paragraph.

Examination of the human body for early stages of superficial spreading melanomas, the most common type, is very easy and can be conducted effectively by almost everyone. One has only to examine moles carefully, noting whether (a) the mole's edges are rough and contain a notch or (b) the color is speckled and contains red, white, or particularly, blue. If either of these conditions are found, examination by a specialist is urgent (very few moles with either of these characteristics are noncancerous).

Figure B.1 shows the specific sites of origin of primary melanoma. (Data from malignant melanoma clinical cooperative group comprising four centers: New York University Hospital, Massachusetts General Hospital, Temple University Hospital, University of California, San Francisco Hospital.) Note the predominant localization of sites of origins to areas of skin that are constantly exposed to sunlight (face and lower legs in females) or to body areas that are intermittently exposed (trunk and legs of males). Also note the virtual absence of melanomas in body areas that are covered (bathing trunk area of males, bathing suit area of females).

REFERENCES

Camina R., A. J. Tuyns, H. Sarrat, C. Quenum, and I. Faye. 1972. Cutaneous cancer in Dakar, *J. Natl. Cancer Inst.* 48(1):33-49.

Climatic Impact Committee. 1975. *Environmental Impact of Stratospheric Flight: Biological and Climatic Effects of Aircraft Emissions in the Stratosphere*, National Academy of Sciences, Washington, D.C.

Cosman, B., S. Heddle, and G. F. Crikelair. 1976. The increasing incidence of melanoma, *Plastic Reconst. Surg.* 57(1):50-56.

Cutler, S. J., M. H. Myers, and S. B. Green. 1975. Trends in survival rates of patients with cancer, *New Engl. J. Med.* 122-124.

Elwood, J. M., and J. A. H. Lee. 1975. Recent data on epidemiology of malignant melanoma, *Seminars Oncol.* 2(2):149-154.

End results in cancer, report no. 4. 1972. DHEW publ. NIH73-272, U.S. Government Printing Office, Washington, D.C.

Fleming, I. D., J. R. Barnawell, P. E. Burlison, and J. S. Rankin. 1975. Skin cancer in black patients, *Cancer* 35(3):600-605.

Lee, J. A. H. 1973. A trend of mortality from primary, malignant tumors of skin, *J. Invest. Dermatol.* 59(6):445-448.

Lee, J. A. H. 1975. Current evidence about the causes of malignant melanoma, *Ariel: Prog. Clin. Cancer* VI:151-161.

Movshovitz, M., and B. Modan. 1973. Role of some exposure in the etiology of malignant melanoma: epidemiologic inference, *J. Natl. Cancer Inst.* 51(3):777-779.

National Center for Health Statistics. 1974. *Vital Statistics of the U.S., 1970*, Volume II, Mortality.

Registrar General's Statistical Review of England and Wales for the Three Years 1968-1970: Supplement on Cancer. 1975. Table E, p. 29.

Scotto, J. 1976. Melanomas of the eye and other noncutaneous sites: epidemiologic aspects, *J. Natl. Cancer Inst.* 56(3):489-491.

Third National Cancer Survey 1969-71: Incidence Data. 1975. National Cancer Institute Monograph 41.

Willis, I., A. Kligman, and J. Epstein. 1973. Effects of long ultraviolet rays on human skin: photoprotective or photoaugmentative?, *J. Invest. Dermatol.* 59(6):416-420.

Appendix C
Biological
Repair Mechanisms

The genetic material of an organism carries the primary store of information from which it must direct the synthesis of its nucleic acid and protein constituents. In the case of cells (in contrast to viruses), this role always falls to DNA. If secondary molecules are damaged, they can always in principle be replaced by new ones synthesized under control of this genetic information; but if the DNA is damaged, there is no other source of directions for producing a substitute. It is therefore understandable why cells should contain specific mechanisms for correcting various kinds of DNA damage.

The types of DNA repair processes known and their characteristics have been summarized in available publications (see, for example, Hanawalt and Setlow, eds., 1975). Briefly, there are at least three different classes: (1) specific enzymatic processes for reversing one particular type of photochemical damage; (2) processes for correcting damage in one strand of a double-stranded DNA by excising it and resynthesizing the faulty structure from the corresponding information in the complementary base sequences of the other strand; and (3) processes that utilize genetic recombination of DNA to create an intact copy of the DNA from two or more copies damaged in different locations.

Type 1 processes are represented principally by photoenzymatic repair, a widespread mechanism that specifically removes the most common photoproduct produced by direct absorption of solar uv in DNA (see Rupert, 1975, for a brief review). This repair process—the first one to be discovered—depends on an enzyme that combines with the damaged site in DNA as its substrate, and in this complex catalyzes a photochemical reaction to restore the normal structure. [A similar mechanism apparently works also in some plant cells to repair RNA (Gordon, 1975).] A second, light-

107

independent example of this type of process is now known, which acts to reverse the major solar uv damage produced in bacterial spore DNA (Munakata and Rupert, 1974; Wang and Rupert, 1976).

LIGHT-INDEPENDENT REPAIR

Processes of the second type, known as "excision repair," depend on a sequence of enzymatic processes. Repair is initiated by a discriminating endonuclease, able to detect an abnormal structure in one DNA strand. This endonuclease produces a single-strand break in the polynucleotide chain adjacent to the damage site, after which excision of the damage by action of an exonuclease, resynthesis by a polymerase, and sealing of the remaining single-strand gap by DNA ligase follow in that order (see Grossman, 1975). This process is remarkably versatile, being able—unlike those of the first type—to repair a wide variety of damages, and there are multiple enzymatic pathways. The discriminating endonucleases isolated from different sources recognize different ranges of nucleic acid alteration as damage requiring correction.

Excision repair cannot function if replication in damaged DNA proceeds beyond the point where a damaged section in one strand is separated from the complementary portion of the other strand. In such cases, processes of the third type could still piece together one good copy from the resulting partial replicas or could make an excision-repairable structure by once again pairing the damaged regions with their complements (see HowardFlanders, 1975). However, little detail is known about the actual operation of this type of process, aside from the fact that recombination processes are involved. It may be artificial to think of the second and third types of repair as completely distinct, since both may involve some of the same enzymes.

Cells of different species depend on the different repair types to different degrees. For general discussion of ultraviolet radiation resistance, however, this is of little consequence. Any type of normal cell employs its own particular combination of repair processes to remove most of the initial DNA damage, the actual extent of repair depending on the amount and kind of damage present and on the precise conditions under which the cell is kept during and after irradiation. The final effect of the radiation depends on how much damage remains unrepaired at the time the cell needs the functions being impeded by it. For most cell types not enough is known to deduce the optimal conditions for fostering repair. Only in the case of some bacteria and rather few higher cells is information on this point available. When an amount of radiation sufficient to override the repair capacities, and leave some net damage, is applied over a few seconds or a few minutes, the exact rate of irradiation seems unimportant. Generally, only the total quantity of damage generated counts, a fact expressed by saying that "reciprocity" holds; i.e., that the product of radiation intensity and time of irradiation determines the effect. The light-independent repairs can overcome a larger fraction of damage when the total quantity initially created is small than they can when this quantity is large. Cells placed under conditions that preclude growth and cell division frequently, but not always, perform excision repair more effectively than cells in good condition for growth and division. (Any such conditions are, of course, difficult to manage without pathological consequences in cells of intact higher plants and animals.) Cells irradiated at very low rates, with the normally inactivating dose spread over many hours, repair themselves much more effectively than cells subjected to acute doses (Harm, 1968). An analogous effect has been noted in skin-cancer induction in mice, where irradiation at lowered rates was less effective than at higher rates (Blum, 1959). The repair systems are evidently adapted to function effectively at lower radiation levels than those often applied in the laboratory.

The significance of dark repair processes in humans is evidenced by the recessively inherited disease xeroderma pigmentosum. Individuals with this condition have a defect in the repair of DNA damaged by uv radiation. They have a predisposition to the development of skin cancer.

PHOTOREPAIR

Photorepair, which requires radiation in either the uv-A or (in some cases) the visible region, can be readily controlled by managing the light environment. The process can utilize longer wavelength sunlight to repair the DNA damage done by short wavelength sunlight (Rupert, 1964), although in the only bacterium in which the matter has been studied, photorepair unaided by other repair systems will not keep up with the damage production in natural sunlight (Harm, 1969). In this bacterium, at the light levels of normal full sun, the rate-determining factor in photorepair is not light intensity but the rate at which the photorepair enzyme can find and attach to damaged sites on the DNA (Harm and Rupert, 1970).

Partial skylight illumination, as through a north window, is highly effective for photorepair, at least in microorganisms, and was in fact the means for Kelner's original discovery of the phenomenon. The effectiveness of light utilization in the repair step—as represented by the product of the molar extinction coefficient of the enzymesubstrate complex times the quantum yield for the repair reaction—is much higher than the corresponding quantity for damage production in DNA (greater than 10^4 liter mol^{-1} cm^{-1} in the first case, compared with ~ 1 liter mol^{-1} cm^{-1} in the second). Moreover, the intensities of the repair-producing wavelengths in daylight are much higher than the intensities of those producing damage. Consequently, it is likely that cells of a mobile animal possessing the photorepair process could recover in well-shaded daylight from damage incurred in full daylight. This would make the net unrepaired damage dependent on the light-seeking behavioral patterns of the animal.

Under optimal circumstances, about 90 percent of the DNA damage is repairable by photorepair. However, the initiation of the process of excision repair will present subsequent photorepair (Patrick and Harm, 1973), so that, at least in many cells, dark repairs shoulder the bulk of the burden. Photorepair is easily demonstrated in cells of species throughout the plant and animal kingdoms, including both plants and animals. It is much weaker in the placental mammals than in lower species, and its possible role in the surface tissues of these creatures is unclear.

REFERENCES

Blum, H. F. 1959. *Carcinogenesis by Ultraviolet Light*, pp. 194-196. Princeton U. Press, Princeton, N.J.

Gordon, M. P. 1975. Photorepair of RNA. In *Molecular Mechanisms for Repair of DNA*, pp. 115-121. P. C. Hanawalt and R. B. Setlow, eds. Plenum Press, New York.

Grossman, L. 1975. Enzymology of excision-repair in bacteria: overview. In *Molecular Mechanisms for Repair of DNA*, pp. 175-182. P. C. Hanawalt and R. B. Setlow, eds. Plenum Press, New York.

Hanawalt, P. D., and R. B. Setlow, eds. (1975) *Molecular Mechanisms for Repair of DNA*. Plenum Press, New York.

Harm, W. 1968. Effects of dose fractionation on ultraviolet survival of *Escherichia coli. Photochem. Photobiol.* 7:73-86.

Harm, W. 1969. Biological determination of the germicidal activity of sunlight. *Rad. Res.* 40:63-69.

Harm, H., and C. S. Rupert, 1970. Analysis of photoenzymatic repair of UV lesions in DNA by single light flashes. VII. Photolysis of enzyme-substrate complexes *in vitro. Mut. Res. 10*:307-318.

Howard-Flanders, P. 1975. Repair by genetic recombination in bacteria: overview. In *Molecular Mechanisms for Repair of DNA*, pp. 265-274. P. C. Hanawalt and R. B. Setlow, eds. Plenum Press, New York.

Munakata, N., and C. S. Rupert, 1974. Dark repair of DNA containing "spore photoproduce" in *Bacillus subtilis. Mol. Gen. Genet.* 130:239-250.

Patrick, M. H., and H. Harm. 1973. Substrate specificity of a bacterial UV endonuclease and the overlap with *in vitro* photoenzymatic repair. *Photochem. Photobiol.* 18:371-386.

Rupert, C. S. 1975. Enzymatic photoreactivation: overview. In *Molecular Mechanisms for Repair of DNA*, pp. 73-87. P. C. Hanawalt and R. B. Setlow, eds. Plenum Press, New York.

Rupert, C. S., 1964. Photoreactivation of ultraviolet damage. *Photophysiology* 2:283-327.

Wang, T. V., and C. S. Rupert. 1976. Evidence for the momomerization of spore-photoproduct to two thymines by the light-independent "spore repair" process in *Bacillus subtilis. Photochem. Photobiol.*, in press.

Appendix D
Sources of
CFM Emissions
and Approaches to
Their Reduction

An understanding of the amounts of CFMs emitted from each use is a prerequisite to any rational approach to the development of a program, if needed, for reduction or control of emissions. In its report of March 1, 1976, entitled "Scientific Review: The Effect of Fluorocarbons on the Concentration of Atmosphere Ozone," the Manufacturing Chemists Association (MCA) presents a compilation of data contributed by essentially all the manufacturers of CFMs worldwide. An earlier (September 1975) report prepared by Arthur D. Little, Incorporated (ADL) for the U.S. Environmental Protection Agency includes more detailed analyses of the uses of CFMs and their estimated emissions within the United States. In order to construct a tabulation of the current rate of emissions from each significant use, we have drawn data from both of these reports and applied several judgmental assumptions to bridge a few gaps not elucidated quantitatively in either report. We have chosen three major categories, namely, aerosols, air conditioning/refrigeration, and plastic foams, as being most logical for organization of this discussion. The assumptions we have made are

(a) That the amount of CFMs used worldwide in the manufacture of open-cell foam in 1975 was 100 million pounds. The MCA report on production and release of CFMs combines the open-cell foam and aerosol applications in a single category. Thus, we have subtracted 100 million pounds from the MCA figure for "aerosols and open-cell foams" and combined it with the MCA figure for "closed-cell foams" to establish values for "aero-

111

sols" and "plastic foams." While we have been unable to find figures that fully support the above value of 100 million pounds, there are statements in the ADL report that make this value appear not unreasonable.

(b) That the breakdown of relative values for the numerous uses listed in Tables IV-2 and IV-5 of the ADL report for the United States applies in similar proportions for worldwide use.

The MCA tabulations, which start from the first year of significant CFM manufacture (1931) indicate use only in the field of air conditioning/refrigeration for 15 years before use in aerosols (1946) and 20 years before use in closed-cell foams (1951). During the past 30 years, use in aerosols has grown very rapidly, far out stripping all other uses. Today 75 percent of all releases of CFMs is from aerosols. It is remarkable that air conditioning/refrigeration, for which CFMs were first developed, now account for only 14 percent of release, and, further, that the ubiquitous and very essential home refrigerator accounts for less than 0.3 percent of the total CFM release.

AEROSOLS

The CFM-propelled aerosol, highly developed in various product designs, has gained a high consumer acceptance and is used daily in one way or another by a very high percentage of the population in nearly all developed countries. Of these many aerosol applications, the personal antiperspirant/deodorant and hair-care uses predominate by a wide margin, accounting for about 56 percent of the total CFM release from all uses.

In nearly every application the aerosol displaced some earlier product that performed reasonably satisfactorily. Today some product to perform the same function is offered in competition with practically every form of CFM-propelled aerosol. Hand application by use of creams, sticks, pads, and pumps are examples. All of these, as well as aerosol applicators, appear to have had very strong advertising support. Why, then, is the CFM-propelled aerosol so widely accepted by the consumer?

Consideration of the following product design factors indicated that several technical attributes of CFMs contribute significantly to their acceptance as propellants.

LIQUEFIED GAS PRESSURIZATION

The CFM, or mixtures of CFM, used in aerosol cans always have boiling points well below room temperature, and the cans are charged with enough liquefied CFM to fill the can with saturated CFM vapor after the entire content of "active ingredient" has been expelled. Thus, the pressure in the can remains constant throughout use, provided the can is kept at a constant room temperature, which is normal. This constancy of pressure contributes importantly to consistency of aerosol particle size, spray pattern, and throw.

BURST EFFECT

For many applications it is desirable to achieve an instant breakup of the droplets leaving the spray nozzle forming many very small particles. CFMs have the ability to dissolve many types of active ingredients or of active ingredients plus an auxiliary solvent. When droplets of active ingredient dissolved in the CFM propellant emerge from the pressurized environment inside the can into the ambient air, the sudden reduction in surrounding pressure allows the CFM to flash into vapor causing the droplets of spray to subdivide into many smaller particles, which instantly become a uniform, gasborne suspension of tiny particles of active ingredient.

The very fine dispersion of active ingredient achieved by the burst effect minimizes the settling rate of the active ingredient in air and thereby increases the persistence of the dispersion of active ingredient in the region adjacent to a discharge of aerosol. While this contributes importantly to the effectiveness of insecticide aerosols, it may be irritating to those users of personal-care aerosols who feel compelled to continue normal breathing immediately following application of the aerosol.

FEEL AND APPEARANCE

By appropriate selection of auxiliary solvents, the speed with which the tiny particles lose all their solvent can be chosen to control the feel and/or appearance of the personal-care product. The antiperspirant/deodorant may be formulated so that the suspension of active ingredient particles arrives at the surfaces to be treated in a dry state, leaving little or no sensation of wetness or stickiness. The hair aerosol may be made to stay fluid on the hair just long enough to coalesce on the surfaces of the hair before drying with an attractive sheen.

HYGIENICS

Ability to use the aerosol without direct contact between the application device and the surface being treated appeals to consumers who may wish to share use of the device with others.

CONVENIENCE

Aerosols are immediately ready for use and require no cleanup after use.

ODORLESS AND STAINLESS

The CFM propellants in high purity do not alter fragrances or leave stains.

PRECISELY METERED APPLICATION

By use of a compound valve and a fixed metering chamber, precisely controlled bursts of aerosol can be discharged as in the administration of certain medications.

Each of the characteristics in the lists above is believed to contribute directly to the widespread consumer use of CFM-propelled aerosols. There are, in addition, the following factors that appeal primarily to the manufacturer and the well-informed consumer.

NONFLAMMABLE

The CFMs are completely nonflammable in any concentration with air. Even blends with significant concentrations of flammable solvents are nonflammable.

LOW ORDER OF TOXICITY

The CFMs with a threshold limit value of 1000 parts per million enjoy one of the least toxic ratings for any chemical.

INERTNESS

The CFMs are unusually inert and present a minimum of problems in achieving long product shelf life.

VERSATILITY

CFMs have a wide range of boiling points and are perfectly compatible with each other. Thus, they can be blended to provide a wide selection of pressures, allowing the product designer to choose a variety of spray characteristics. Solubility can also be altered by varying the propellant blend, thus making it possible to achieve a special effect, such as the release of certain foams.

DENSITY

The high density of liquefied CFMs facilitates maintaining suspension of certain insoluble active ingredients.

ALTERNATES TO CFMS

HYDROCARBONS

Except for flammability and density, hydrocarbons generally qualify under all the characteristics noted above. Furthermore, they are lower in cost and therefore provide a strong incentive for the manufacturer to choose hydrocarbons in preference to CFMs. However, the high flammability of hydrocarbons seriously limits their acceptability for general consumer use except where ample ventilation and freedom from sparks can be assured.

COMPRESSED GASES

Compressed gases, such as nitrogen, nitrous oxide, or carbon dioxide, are used as aerosol propellants but suffer in comparison with CFMs because the propellant pressure declines as the charge of active ingredient is consumed, leading to a lack of uniform performance over the life of the product. Insofar as it is possible to obtain a solvent for the propellant gas that is compatible with the active ingredient, this loss of pressure effect may be ameliorated but not completely overcome.

MECHANICAL SPRAY PUMPS

Finger-actuated spray pumps and hand-actuated flit-guns were offered to the consumer long before the development of CFM-propelled aerosols. The current climate of controversy over the release of CFMs has spurred renewed effort in the further development of pump delivery systems. Achievement of a uniform dispersion of tiny spray particles thrown a suitable distance with a finger- or hand-actuated pump is far from simple. Conceivably, a high-pressure system, driven electrically, could approach the performance, but not the convenience, of the aerosol can.

Some of the present applications of aerosols, other than personal care, do not require the fineness of particle achievable with aerosols. Here, finger-actuated pumps enjoy significant acceptance and are reasonable candidates for much wider acceptance. However, it is well to note some of the problems that tend to plague pump systems.

Leakage   The comparative complexity of seals in pump systems makes them more susceptible to leakage of active ingredient during storage and handling. There is also the possibility of air leaking into the active ingredient chamber, with the possible result that the active ingredient will be prematurely degraded.

Spray Characteristics   As implied above, the spray from pump systems is usually in the form of comparatively coarse droplets and is apt to be nonuniform. Under some circumstances dripping or solid streaming may result in inconvenience and may even damage objects outside the target area.

Availability   Limited manufacturing capability for pumps now exists. A broad-scale displacement of aerosols with pump systems would require significant expansion of pump manufacture and adjustments in related portions of the industry.

AIR CONDITIONING/REFRIGERATION

In view of the fact that there are about 69 million home appliance refrigerators in use, as compared with about 40 million automobile air conditioners, explanation of the striking disparity in their contribution to emissions, as indicated in Table D.1, is in order. The home appliance refrigerator is a hermetically sealed unit, whereas the automobile air conditioner is not. Typically, home refrigerators operate for their useful life without requiring any service to the refrigerant circuit. On the other hand, the flexible elastomer hoses used under the automobile hood for conducting refrigerant to and from the engine-mounted compressor in the typical automobile air conditioner are an imperfect vapor barrier, which allows slow loss of the refrigerant by diffusion through the hose wall. The rate of diffusion increases with ambient temperature. Thus, the frequence at which replacement of the refrigerant charge becomes necessary would be expected to vary, depending partly upon thermal exposure. Another factor in the design of automobile air conditioners that contributes to loss of refrigerant is the rotary seal on the compressor drive shaft. Although highly developed and capable of functioning very well for many years, it may fail prematurely if not periodically "exercised" to maintain an oil film between its precision faces. It is not clear how much loss of CFMs would be saved if motorists appropriately "exercised" their air conditioners in the winter. Also, the hose connections and several other static seal joints used to confine the refrigerant are mechanical in nature and may loosen under severe usage. The combined effect of these three factors is a loss in refrigeration capacity of the system, and, thus, a requirement for service, often more than once during the life of the system.

TABLE D.1 Worldwide Releases of cfms (Millions of Pounds)[a]

AEROSOLS 1115.1 (74.5%)	Personal Care	Antiperspirants/ deodorants	458.4
		Hair care	401.5
		Medicinal	37.3
		Fragrances	2.3
		Shave lathers	0.9
		Others	34.4
	Household	Room deodorants	17.7
		Cleaners	9.6
		Laundry products	23.4
		Waxes and polishes	9.2
		Others	9.2
	Miscellaneous	Insecticides	33.3
		Coatings	22.9
		Industrial	39.0
		Automotive	8.0
		Vet. and pet.	2.3
		Others	5.7
AIR CONDITIONING/ REFRIGERATION 204.7 (13.7%)		Mobile A/C	89.8
		Chillers	42.9
		Food store	33.1
		Beverage coolers	5.8
		Home refrigerators and freezers	5.8
		Miscellaneous	27.3
PLASTIC FOAMS 176.5 (11.8%)		Open cell	100.0
		Closed cell	76.5
			1496.3

[a]Based on the annual incremental releases of F-11 and F-12 indicated for 1975 in the MCA report and the detailed percentage analysis by uses in the United States for 1973 in the ADL report. Russia and Eastern Europe, which have not reported production or release data (about 11 percent of total), not included.

The available figures on consumption of F-12 in the servicing of automobile air-conditioning systems suggest that substantially more refrigerant is used than would be expected from the leakages enumerated above. This leads to the belief that present servicing methods involve venting much more refrigerant than the amount that leaks. Observation of the methods commonly used in servicing support this conclusion. Generally accepted servicing practices are influenced by the concept that the entire refrigerant charge should be released and replaced with the prescribed measured quantity of new refrigerant if there is any evidence, suspected or real, that moisture or air has gotten into the refrigerant circuit. This is perhaps the most reliable way of establishing the correct charge in the system, because it is not easy to determine exactly how much charge remains in a system after an unknown amount of leakage has occurred. Since the F-12 may be present in both its liquid and vapor phase, simple pressure measurement will not suffice to determine the amount of F-12 in the system. Practically all systems are now equipped with a transparent sight glass in the liquid passage, leaving the "receiver-drier," which enables the serviceman to make a quick, qualitative visual check on the adequacy of the refrigerant charge. The appearance of bubbles while the system is in operation is usually an indication of insufficient charge. If only a "topping-off" of the charge is needed, this can be accomplished with very little CFM emission simply by adding refrigerant a little beyond (say, 4 ounces) the point at which bubbles disappear from the sight glass.

The reasons for releasing the F-12 from an automobile air conditioner are not limited to malfunction of the air-conditioner system. Often, the need for servicing some other part of the automobile may necessitate removing a component of the air conditioner in order to gain access to the part to be serviced. For example, replacement of a radiator usually requires removal of the air-conditioner condenser, which is mounted close in front of the radiator core. In such situations it is standard practice to release the refrigerant charge just as one would release the air from a tire before dismounting it.

By use of a small compressor connected to an air-cooled condenser coil and pressure receiver tank, any service mechanic competent to recharge the refrigerant circuit in an automobile air conditioner could easily retain the old charge instead of venting it to the atmosphere. In applying such a recovery system it would be necessary to avoid admitting "noncondensable" air into the system. Thus, when "pumping out" the refrigerant circuit to less than atmospheric pressure, it would be important to avoid in-leakage of air because of its adverse elevation of pressure in the used refrigerant accumulation tank. Except as such a buildup of pressure might require premature return of the accumulation tank to a reprocessing center, used F-12 would be accumulated in the tank until its gross weight reached a prescribed level for return to the reprocessing center, where its contents would be repurified and packaged for reuse. Judging from the consumption figures at hand, it appears that a reduction in the total CFM emissions from refrigeration applications of about 24 percent (3.3 percent of total emissions) might be accomplished by universal adoption of procedures for retention and recycling of F-12 by those who service automobile air conditioners.

Application of this same charge-retention procedure and equipment to recover F-12 from air-conditioning systems of automobiles just prior to being scrapped might give an additional 14 percent reduction (1.9 percent of total emissions). In other words, over one third (one twentieth of overall total emissions) of the total emissions from the field of air conditioning and refrigeration might be stopped by instituting a universal procedure for retention and recycling of F-12 during the servicing and scrapping of automobiles. This is a step that could be implemented on the basis of well-known technology without requiring expensive or time-consuming technical developments.

In view of the fact that F-12 is universally used in automobile air conditioners and the belief that F-22 emissions would be much less detrimental to the ozone shield than emissions of F-12, the possibility of substituting F-22 for F-12 in automotive air conditioning is frequently brought up. The prospects for success in making a simple substitution of F-22 for F-12 in existing systems are very poor. The thermodynamic properties of F-22 require operation at significantly higher pressures and reduced volumetric flows. Attempts to use F-22 in an automotive air-conditioning system nicely attuned to operation on F-12 would very likely lead to early failure because of excessive loads on the compressor bearings and clutch. It is doubtful that the controls would function satisfactorily without readjustment or replacement. Loss of refrigerant by diffusion through the wall of elastomer hoses would increase. In short, it would be necessary to replace the entire system with one designed for operation on F-22 in order to achieve fully satisfactory operation. A further complication in making such a substitution would be the necessity of increasing the plant capacity for manufacturing F-22 in support of such a large change in demand for F-22.

Table D.1 shows the large commercial chillers as an important source of emissions. These chillers, including both large centrifugal and large reciprocating compressor types, are used extensively for central chilling of water that is circulated throughout large buildings, such as hospitals, hotels, and office buildings, to provide cooling to each room. Typically, these systems are so large that refrigerant lines are connected after the major components are erected, and achievement of leak-tight systems is considerably dependent on the competence and integrity of the installer, often working under somewhat adverse conditions. Both during initial erection and subsequent servicing of the system, rigorous attention to workmanship and well-planned application of appropriate leak detection techniques and instrumentation are required to avoid leakage losses. Here again, any present practices, which have condoned venting refrigerant during service activities, should be changed to routine use of procedures and equipment for retention and recycling of any portion of the refrigerant charge removed from the system.

Table D.1 also indicates that food-store refrigeration is an important source of emissions. The observations concerning erection and servicing of large chillers apply similarly to this type of use.

From the foregoing considerations it is concluded that upward to 85 percent (11.6 percent of overall total) of the total emissions from the air-conditioning and refrigeration application of CFMs could be eliminated by application of appropriate methods for prevention of leakage, venting during servicing, and removal of refrigerant charge prior to scrapping of systems.

Long-term objectives might well include development of air-cycle refrigeration equipment for air conditioning of private automobiles. The prime obstacle to near-term application is the lack of rotary compressor/expanders with high enough efficiency in the appropriate capacity range. At least two existing developmental programs have reasonable prospect of yielding suitable compressor/expanders, provided adequate support is maintained. Presumably, the methods appropriate to servicing air-cycle equipment would fit easily with methods typical of the automotive field.

In many home and public building air-conditioning applications the water-lithium bromide absorption system is a technically proven option. Thoughtful integration of this type of system into the overall system for energy utilization would yield significant reduction in fuel consumption. This brings up the intriguing possibility that with continuing rise in fuel cost the greater initial cost of an absorption system compared with a vapor compression system may not be a deterrent to its selection.

Because of the fact that water is the refrigerant in the waterlithium bromide absorption system, use of this approach to refrigeration is limited to temperatures above the freezing point of water. For lower temperatures, a practical alternate to use of a CFM refrigerant is the use of ammonia. Ammonia's toxicity, flammability, and corrosive effect on some metals would, of course, make it necessary to design a system to accommodate these disadvantages. The toxicity and flammability characteristics would usually require that the ammonia circuit be isolated so that refrigerant leakage would not become a hazard to people or property. Isolation of the primary refrigeration circuit would usually require use of at least one secondary circuit of low-temperature heat-transfer fluid such as brine. The technology of this type of system is well established, but, obviously, the design adaptability is apt to be poor unless detailed plans can be laid out at the time the architectural plans are under way.

PLASTIC FOAMS

The present state of plastic foam technology and product perfection represents years of intensive research and development, as well as extensive investment in specialized manufacturing facilities. Many different blowing agents, or combinations of blowing agents, are used. CFMs are used in varying degrees, depending importantly on the particular foam characteristics desired for specific applications, as well as the CFM favorable characteristics in terms of nonflammability, low toxicity, freedom from odor, noncorrosiveness, and compatibility with many polymers.

Rigid polyurethane foam, with its cells essentially permanently charged with CFM, is a unique material in terms of good structural characteristics coupled with low thermal conductivity. It is the material that makes the "thin-wall" home refrigerator practical. The living space gain through availability of thin-wall refrigerators is clearly an important consideration for many users. While other approaches to thin-wall construction are technically possible, none has been demonstrated to have the overall practicability of the rigid, closed-cell, CFM-charged foam.

The extent to which use of CFMs in producing open-cell foams could be displaced with substitute blowing agents involves both questions relating to product characteristics in terms of consumer acceptance and acceptability of substitute blowing agents in the manufacturing environment. In particular, the high flammability of several possible alternate blowing agents presents very serious problems to any manufacturer wishing to use them. At best, extensive replacement of CFMs as blowing agents in the manufacture of plastic foams is likely to involve protracted research and development, as well as extensive new plant investment and accompanying industrial dislocation.

Appendix E
Preventive Measures
Related to Melanoma

/

PREVENTIVE MEASURES

With melanomas presently providing about 1 percent of all cancer deaths, and with a relatively clear understanding of both causation and prevention, it would not be appropriate to discuss the increased risk from increased DUV without discussing the possible decreases in risk that an effective practice of preventive measures could contribute. Since most of the effective measures require active cooperation by the persons involved, there are real difficulties in the way of an effective program, but the potential—and even the plausible—reductions in unnecessary deaths make the outlining and con- sideration of a program of preventive measures essential.

Such a program would involve three phases:

—Identification of persons of higher susceptibility (at worst, types I and II skins, see below).
—Adequate protective measures, continued through life, by these more sen- sitive individuals.
—Regular educated self-examination of these individuals, in order to dis- cover skin cancers in their earlier stages when treatment (surgery) is effective. (The amount of potential benefit is still a controversial issue. There is no information from well-designed observational or experimental studies to justify a definitive conclusion that early detection and treatment

changes the natural history of disease. However, medical oncologists believe that clinical experience points to prolonged survival, early detection and treatment. The major point of debate is the proportion of melanomas whose consequences can be effected by such measures.)

That a significant fraction—conceivably as much as one third—of our population should have to plan to protect themselves, throughout their whole lives, from the ill effects of excessive sunlight will be surprising to many—and shocking to some. Yet, until and unless we find better ways to focus on a smaller group of susceptibles, nothing less can do very much to reduce skin-cancer incidence.

The measures that need to be taken include

—Avoidance and reduction of exposure
—Daily use of effective sun-screen preparations (most generally used suntan preparations are ineffective for this purpose).

Most people who are more highly susceptible are likely to need to use effective sun screens, although some may be able to reduce their exposure so drastically that sun screens are not essential.

SKIN TYPES

Different persons are very differently susceptible to sun-induced skin damage. Personal history of sunburning and/or suntanning following the first 30-minute exposure of the summer to sun is ordinarily sufficient to classify whites in four types of grades of reactivity. (See Table E.1.)

TABLE E.1 Working Classification of Sun-Reactive Skin Types Used in Clinical Practice[a]

Skin Type[b]	Skin Reactions[c] to First 30-Minute Exposure of the Summer
I	Always burn, never tan
II	Usually burn, tan below average (with difficulty)
III	Sometimes mild burn, tan about average
IV	Rarely burn, tan above average (with ease)

[a]More precise and sensitive classification will be needed as knowledge evolves.

[b]Type I and Type II persons often have light skin color, blue eyes, may have red scalp hair, and may or may not have freckling. However, some persons with dark brown hair and blue or green eyes have Type I and Type II sun reactions.

[c]At age 12-40 years.

SUNSCREENS

Easily applied sun lotions containing about 5 percent of para-aminobenzoic acid (PABA) in an alcohol-water mixture are now commercially available. In carefully controlled studies under rigorous field conditions (Pathak et al., 1969; Fitzpatrick et al., 1974), a single application of these formulations provided day-long protection against the ill effects of DUV, even under conditions of profuse sweating. They were not, however, effective after prolonged swimming, so that an additional application would have to follow each extended swim.

For persons who object to the drying effects of the alcohol, or for the few percent of people who have contact sensitivity to PABA, ointment-based preparations containing cinnamate and benzophenone derivatives (e.g., ethylhexyl-para-methoxycinnamate and 2-hydroxy-40methoxybenzophenone, supplemented by 2-phenylbenzimidazole sulfonic acid) are available and have proved equally effective.

These invisible sun-screen lotions should be used daily in the morning, except on dark, cloudy days. As noted above reapplication may be needed after prolonged bathing.

They should be applied as follows:

—To the faces and hands of all susceptibles.
—To the legs of girls and women
—To the legs of males whenever shorts are worn
—To the upper body, arms and forearms, whenever, as in summer, short-sleeved or incompletely opaque shirts or blouses are to be worn.

The International Congress of Photobiology is (1976) in the process of formulating standardized protocols for qualifying the effectiveness of uv-B-absorbing sun screens. Thus, regulatory control, either through labeling requirements or proof-of-effectiveness regulation may soon be available.

FURTHER POINTS

Sun damage to skin is often equated with acute sunburn, but sun-induced skin damage is cumulative and irreversible, and repetitive short exposures with no visible sunburn can lead to skin damage. The following points are important:

1. DUV is most intense between 11:00 a.m. and 3:00 p.m. (daylight time)—accordingly susceptibles will do well to schedule exposures before 11 or after 3.

2. Some clothing (specifically that through which underclothing is identifiable) transmits more than 50 percent of DUV—thus susceptibles will often need to apply sun screens to their upper body.

3. DUV intensities in the shade are often 50 percent of those in bright sun.

4. 70 to 80 percent of the bright-sun DUV reaches the ground on bright but hazy days with clouds.

5. A sand background increases the DUV by 25 percent; snow by 100 percent.

REFERENCES

Fitzpatrick, T. B., M. A. Pathak, and J. A. Parrish. 1974. Protection of human skin against the effects of sunburn ultraviolet (290-320 nm). In *Sunlight and Man*, pp. 751-765. M. A. Pathak *et al.*, eds., U. of Tokyo Press.

Pathak, M. A., T. B. Fitzpatrick, and E. Frenk. 1969. Evaluation of topical agents that prevent sunburn—superiority of para-aminobenzoic acid and its ester in ethyl alcohol, *New Engl. J. Med.* 280: 1459-1463.